GW01034462

ISBN 978-1-5285-2213-7
PIBN 10902263

FB &c Ltd, Dalton House, 60 Windsor Avenue, London, SW19 2RR.
Company number 08720141. Registered in England and Wales.

For support please visit www.forgottenbooks.com

POCKET-BOOK

ON

COMPOUND-ENGINES,

BY

nicholas practer

N. P. BURGH,

MARINE CONSULTING ENGINEER, PAST PRESIDENT OF THE INSTITUTION OF MARINE ENGINEERS,

AND

AUTHOR OF THE FOLLOWING WORKS:—

Slide Valve.
Indicator Diagram.
Marine Engineering
Screw Propulsion

Link-Motion.
Marine Compound-Engines.
Details of Screw Propellors.
Pocket-Book of Practical Rules.

Condensation of Steam
Principles of Engineering.
Boilers and Boiler Making.
Science of Steam.

SECOND EDITION REVISED AND ENLARGED.

N. P. BURGH, LONDON.

1887.

IMPORTANT NOTICE

TO STEAM POWER USERS!

Registered Trade Mark.

Do not have your Valve faces and Cylinders cut or grooved, or hot and cut bearings (which take years if ever in some cases to work up to a face again). This Oil does not attack India Rubber Valves of Air Pumps, and stands up to 150° Fahrenheit. Having examined Engines that have used my Oils continually for the past eight years, is sufficient guarantee of their superior quality.

SEND TRIAL ORDER TO

JOHN ETHERINGTON, M.I.M.E.

Consulting & Inspecting Engineer,

SOLE MANUFACTURER & REFINER OF CYLINDRINE LUBRICATING OILS,

39A, KING WILLIAM ST., LONDON BRIDGE, E.C.

Telegraphic Cipher Address—"ETHERINGTON, LONDON." Works—U.S.A.

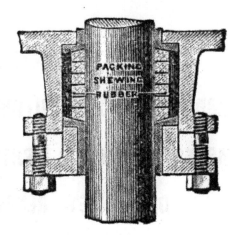

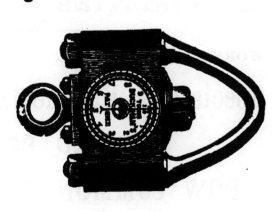

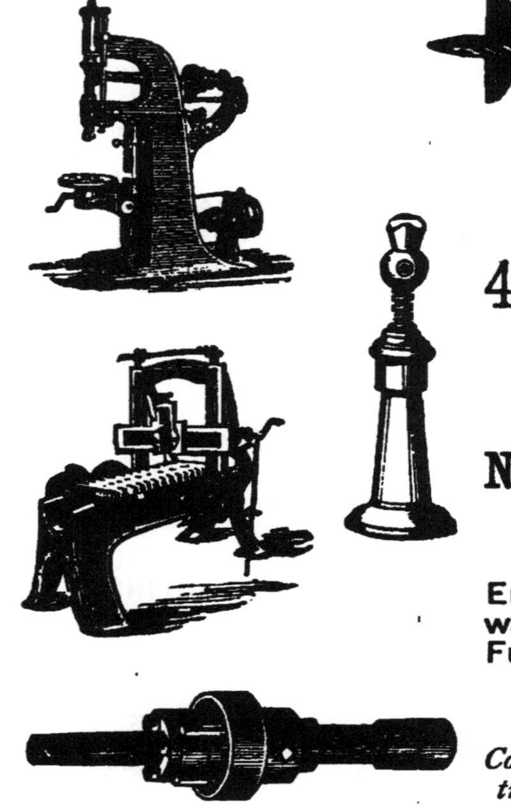

No. 3.
No. 6.

CONTENTS OF THE NEW MATTER IN THE SECOND EDITION

At the end of the Index is a List of the Prepaid Subscribers.

INTRODUCTION

To the Second Edition of this Work.

In order to make this Edition a reliable guide for Engineers at this stage of advancement in their profession and practice, I have treated, in the additional matter, on Locomotives, and Marine Compound Engines. I also have dilated on closed stoke-holes, forced, induced and natural draughts, in a tabulated form. On liquid fuel I have touched upon, but not as much as I should have wished, because my space herein has been confined principally to the engine rather than the boiler As regards locomotives, Webb and Worsdell have contributed much reliable information, while the results shown are much alike. The arrangements are entirely apart in design, and positions for the cylinders; as for instance; Webb uses two high-pressure cylinders and one low-pressure cylinder, and two separate driving shafts; but Worsdell has only two cylinders, high and low pressure, and one driving shaft. Webb has therefore two side powers and a central third power, but Worsdell's powers are side by side, acting together. Perhaps, a better result might be obtained by a pair of compound cylinders, end to end, in "tandem" fashion, the pistons actuating two crank pins on the one shaft.

With Stationary Engines, "we are much as we were." Galloway has introduced an arrangement of two cylinders, angularly situated, the high above the low-pressure cylinder, and both pistons actuate one crank pin. "Tandem" engines for horizontals are still in use, and equally so are the side-by-side arrangements, for two crank pins, or the beam.

Taking Marine Engines next into notice—and thanks to the firms contributing the information recorded herein,—much advance has been made within the last few years in the way of vertical "tandem" engines and the triple system, the latter being one high and two low-pressure cylinders for the use of the one volume of steam, from the admission to the final exhaustion—is now the "fashion" amongst the best makers, and doubtless, in due course, with higher pressures, say, 250 to 300 lbs. on the square inch, the quadruple arrangement must follow. It being understood now, as I stated in the year 1876, that the amount of units of heat in the steam is the vital power obtainable. I may as well add, that the pressures most in use now range from 90 to 150 lbs. per square inch.

The use of receivers are now becoming ignored,

because engineers are learning that steam must be "kept alive" or in motion of traverse whilst at work, from the fact, that when a volume of steam is "cut off" from the boiler it has no more effect than what it contains, and if it stops or sleeps, it has to move or "wake up" again; and this "waking up" absorbs a certain amount of the elastic force of the steam, and thereby reduces its power.

Whilst on this subject I may as well say a word or two about the pressure of the steam at the point of exhaustion. I am afraid it has been much overlooked that the "exhaust" can be at too low a pressure. I am afraid, too, that we engineers, in our hurry, have practiced as our predecessors much on this matter, *i.e.*—"exhaust, as low as you can for the sake of the vacuum." To my mind, there is a graver consideration "in the way" of the friction of the engine; as for example, if the steam is exhausted at a pressure below the power required to overcome the friction, there must be a loss, inasmuch that the steam-effort operating before the exhausting point, has to complete the stroke of the piston at a loss of power, or so much taken out of it thereby, or what may be termed momentum of piston action. I owe a deal of this argument to T. Russell Crampton, the pioneer of the Express

Locomotive, and have put it in a practical form further on.

Referring next to the steam or slide-valves for Marine Engines, we are drifting back into the use of the hollow and solid piston valves, *i.e.*—valves inside the other or separate, but as pistons all the same.

With reference to Triple Marine Compound Engines and Closed stoke-holes, Parker, Wyllie and Sennett, at the Inst's. of Nav. Arct's and Mech. Eng'rs. have given the best information I know of on the subject, and to them be the credit.

Next comes Martin, on induced Draughts, *i.e.* —Open Stoke-hole and a fan in the smoke-box or uptake, he contends this is an advantage over the closed system and forced draughts, and even without the former.

Passing to Liquid Fuel, Admiral Selwyn has given me the information, and from the fact that he has laboured in the cause for many years, I take it to be the most reliable I could obtain.

To the pre-paid subscribers to this Edition, I thank you indeed, "one and all."

N. P. BURGH.

London, December, 1886.

CHAPTER A.

TRIPLE-COMPOUND ENGINES.

The use of three cylinders to expand one volume of steam has at last come into practice in the Mercantile Navy and with good results, in comparison with the two cylinders compound engine.

The main advantage derived by the compound system, is to allow time and space for the heat in the steam to expand the water contained therein, it being well known (as stated on page 4) that the constituents of steam are "heat, air and water."

The presence of the "heat" commences the moment evaporation occurs. The amount of air is due to the quantity of water in the boiler, and the relative amount of feed water required that is due to the eflux of the steam or volume consumed by the engine.

The amount of water relative to the volume of steam therefrom is known by the formula $W = \frac{S}{V}$ where W equals the amount of water, S equals the amount of steam, and V equals the relative volume of the steam compared with the water; the steam tables on pages 196 to 204 explain this, from a pressure of 1 lb. to 1,500 lbs. on the square inch, and the formula above is explained on page 205.

In consideration of this we will explain the practical utility of the table referred to, the headings of which are Pressure, Sensible Temperature, Total Heat, Weight, and the comparative relation of the water to the whole constituents.

Suppose now we arrange to calculate for a triple compound engine say of 4780 I.H.P., and the pressure of the steam 160 lbs. on the square inch in the boiler, and initially 150 in the engine high pressure cylinder. Our next matter is to determine the volume of steam in proportion to the speed of the piston, that relates to the indicated horse-power named, and to do this we must refer to the amount of *heat* required, then comes the value of our table of constants on page 176.

Then as the constant for 4780 equals 2·104 lbs. deduced in the following manner:—constant 2·078 refers to 4330 I.H.P.; and the constant 2·130 refers to 5230 I. H. P., then the mean of the powers equals 4780 and the mean of the two constants equals 2·104, thus we have a starting point.

We turn now to page 70, there the formula as therein expressed, states that $U = \frac{P}{C}$ when U equals units of heat required in relation to the indicated horse power P, and the relative constant value equals C, or heat constant.

Then 4780 divided by 2·104 equals 2271, as the units of heat required to produce 4780 I.H.P. We may as well state at once that the table of the constants on page 176, recognise the frictional power and the speed of the piston; both are purposely given comparatively high and low, to allow for contingencies, *i.e.* on the "safe side."

Now as to the value or application of the 2271. Then comes the consideration of the relative temperature of the steam herein known as 150 lbs. on the square inch, and on page 202 we find it recorded as 366° Fahr.

Learning next from the formula on page 70, that as the units of heat belong directly to the relative temperature of the sensible heat, the former must be divided by the latter; then as that produces the thermal quantity or weight of the steam, thus we know therefore the amount of the steam from the tables corresponding to the known pressure, hence this:—$B = \frac{U}{T}$ where B equals the weight of the volume, U the units of heat, and T the sensible temperature, and putting it into figures thus—2271 divided by 336 results in producing 6·204918, as the weight in lbs. of the volume of steam required in this case.

We know now two facts from this latter portion of the formula—that is, the amounts of water and steam relatively required— see also page 62, and in fact the whole of chapter VII.

Having the weight of the steam before us we next put it into capacity or cubical quantity, and as our steam table gives the weight in cubic feet, we divide the total weight by the weight of one cubic foot or capacity of the relative volume; therefore let $F = \frac{B}{S}$ known as F, to equal the cubical contents of the said volume, and S the weight of one cubic foot of steam correspondingly.

The figures are therefore 6·204918, divided by ·3714, makes the sum of 16·7, as the cubical contents of the supply steam in feet, and that sum 16·7 multiplied by 1728 equals the contents in cubic inches 28857·6. The stroke of the piston to be four feet as an agreed convention.

Next we deal with the area of the high pressure cylinder, because that is the main agent. From practical experience we prefer to recommend a "cut-off" of *one-third* of the piston's stroke for general full-power steaming, and of course we must take the maximum in this case, therefore the length of the cut off in this case is 1·333 lineal feet, or 15·996 lineal inches.

Now we go back to the cubical contents of the steam, which it will be remembered is 28857·6 inches. Then if we take the length of "cut-off" as—say 16 inches—and divide 28857·6 by 16, we have 1803·6 as the area of the high pressure cylinder which is about or within an ace of 48 inches in diameter.

The one volume of steam operates for one and half revolution, or for three strokes of the respective pistons, hence there are three mean pressures to be considered, or the theoretical diagram expressed longitudinally, equals the *full* line and the *expansion* curve and will represent 12 feet in this instance.

From the fact therefore that the indicator figures from 2 or 3 cylinders, or more, of a compound engine, is as but one diagram or figure from one cylinder, inasmuch that the compound engine with 2 or more cylinders, is as but one engine practically, and the only reason for introducing the multiple system is to allow time and reduce continuous space for the continuous action of the steam on the crank pins.

Now for the relative areas of the three cylinders separately. We have given already the ratio of supply in the high pressure cylinder as 1 to 3, so therefore the mean pressure is 100 lbs. on the square inch—from the accepted fact that the initial pressure divided by the ratio of supply equals the pressure at the point of exhaustion, and those sums added together and their mean taken equals the mean pressure for the entire stroke of the piston. Therefore this axiom:—*As the mean pressures are relative to the areas of their respective pistons to produce the piston, power, so are the pistons, areas relative to each other* It follows, therefore, that the area multiplied by the mean pressure equals the force of piston-power.

In the present instance the first piston-power is 180360, which, divided by the mean pressure in cylinder No. 2, equals 4874 area, or 78¾ inches diameter; and the same 180360, divided by the mean pressure in cylinder No 3, equals 8588 area, or 104⅝ inches diameter. The sums of the mean pressures are known by the formula already quoted:—perhaps we may as well state that the ratio of supply steam in cylinder No. 1, is $\frac{1}{3}$; in No. 2, $\frac{1}{2}$; and in No. 3, $\frac{2}{3}$; and the final exhaust is 17 lbs on the square inch,—but in actual practice deduct $\frac{1}{3}$, or say 12 lbs. on the square inch, for the exhausting pressure of the steam will cover the frictional power.

The following are the particulars therefore arrived at by this explanation, and formulæ for the Triple Compound Marine Engine of Modern Practice.

Proportions of a Modern Triple-Compound Marine Engine, from the Formulæ herein.

Total Indicated horse power		4780
Pressure of initial steam in high pressure Cylinder		150 lbs.
Constant for units of heat		2·104
Amount of units of heat per stroke ..		2271
Length of first cut-off or initial supply in inches		16
Stroke of piston in inches		48
Area of high pressure cylinder, ..	No. 1	1803
Area of intermediate pressure cylinder	No. 2	4874
Area of low pressure cylinder ..	No. 3	8588
Indicated pressure initial	No. 1	150 lbs.
,, ,, ,, ..	No. 2	50 ,,
..	No. 3	25 ,,

Indicated pressure exhaust	.. No. 1	50	lbs.
,, ,, ,,	.. No. 2	25	,,
,, ,, ,,	.. No. 3	17	,,
Indicated pressure mean	.. No. 1	100	,,
,, ,, ,,	.. No. 2	37,5	,,
,, ,, ,,	.. No. 3	21	,,

Thus far we are on safe ground because we have been dealing with recognised facts, but we now come to the present unknown quantity, *i.e.*, "the speed of the piston in feet per minute." This has been a difficult problem with engineers at the outset in their calculations to obtain the numeral of the indicated horse-power, and we claim to have "solved that problem," as explained on page 154, and from that formulæ the present data is as follows:—

Indicated horse-power	P =	4780.
Theoretical mean pressure ..	Z =	52·666·
Constant value	C =	2·104.
Maximum power	Y =	6309600.
Collective pistons, area	A =	15265.
Power in foot pounds	V =	157740000.
Multiplier	K =	33000.
Speed of piston in feet per minute	S =	420.

We may remark here that the piston speed is low, but in practice it will be higher, as will also be the I.H.P. in proportion; but there must be a margin with all calculations that lead to a successful issue, which in this case is a certainty.—BURGH.

ANGLES OF THE CRANKS FOR TRIPLE-COMPOUND ENGINES.

This is a subject requiring some consideration in this way:—the forces on the crank pin at any number of points on the circular path can be known by the parallelogram of forces applied to each point, and the sums to scale set off from a straight representing the circumference of the said path; we shall then have by joining the limits of the "set-offs" a cirvular or zigzag line representing the forces on the crank pin comparatively for the two strokes or one revolution. We have in practice tested this for some time and know of no better method.

The angles of the cranks for three pistons acting on separate pins are usually equidistant on the circle, but we believe and indeed recommend, that the two cranks for the low pressed pistons should be in a line with each other, or the crank pins opposite each other, while the high pressed piston's crank, should be at right angles with the other two. Our reason for stating this is that the best or most effective effort of the crank pins is thereby equally maintained by the two low pressed, while the high pressed is at ts worst position or in a line with piston rod.—BURGH.

DIFFERENCE BETWEEN EXPANSIVE AND COMPOUND ENGINES.

A practical definition of the differences between a single cylinder "expansive" and two or three cylinders "compound."

1.—Draw a circle denoting the path of the crank pin.

2.—Imagine the piston rod, connecting rod, and crank to be on one line.

3.—Set off from each side of the circle inwards one-third of its diameter.

4.—Project lines from those points at right angles to the crank line, bisecting the circle.

5.—Mark each bisection A to B and C to D respectively on the circle.

6.—Then, A to B is the effective effort of the crank pin from the first volume of steam, and from C to D is the effective effort of the crank pin from the second volume of steam, both being separate and new volumes, entering each end of the cylinder for an expansive engine.

7.—Then, A to B is as before, but C to D belongs to the same volume instead of a second new volume, hence two efforts instead of one from one volume, with a two-cylinder compound engine.

8.—Then, A to B and C to D, as before, with two more points on the same circle, but with a separate crank line for a three-cylinder compound engine, hence three efforts from one volume of steam.—BURGH.

RELATIVE EXHAUST PRESSURE TO THE FRICTIONAL POWER OF THE ENGINE.

As stated in the introduction, this subject has been much overlooked, and it is our purpose now to explain its importance.

Taking the indicator diagram as the leading portion, we notice first the amount of the steam line (or "length of cut off") then that sum or quantity is the width of the parallelogram of the volume of steam used per stroke of the piston, this represents also the power attainable from the supply and expansion which we will designate as supply-power, and expansion-power. The "supply-power" is therefore due to the contents of the volume and its continuous pressure, while the "expansion-power" is due to the elasticity of the volume and the relative reducing pressure, but the "frictional-power" is constant.

Assume next the frictional power is equal to 6 to 8 lbs. on the square inch of the piston's area, next follows this axiom, viz., *the pressure at the point of exhaustion must exceed the pressure absorbed by friction or inertia.*

Next to be noticed is the various parts of the engine that cause the frictional power. The air and circulating pumps are the main agents in the affair, and their resistance or power absorbed is easily obtained from indicator diagrams of the force they exert. In fact the power of a pump is always obtainable from this same formula as for the power of the steam engine.

Taking the entire friction of the engine and its pumps in round numbers it amounts to $\frac{1}{8}$th to $\frac{1}{7}$th of the indicative power, as the resistance in practice, that is, when we talk of the "indicated horse power developed," we must remember that what drove the ship is that sum minus the "frictional power," and the remainder is the effective working-ship-power—to coin an expression.

Air pump power—is due to the amount of steam condensed, and the quantity can be always known from the "relative volume of the steam compared with the water from which it was raised," as explained on page 205; it being always a fact that the air pump is a water pump also.

Circulating pump power—is due to the amount of water required in proportion to the area of the tube surface (inside or outside), and the height that the water is discharged—the table on page 168, refers to this principally.

The abolition of the air-pump is only a matter of a short period (as, has been stated before in another place), inasmuch that vacuum of 12 to 14 lbs. on the square inch is not to be compared with the power saved by its absence, and the thereby increased temperature of the feed water, &c.—See page 129, Chap. xi.

The abolition of the air pump also reduces the cubical contents of the circulating pump, and equally therefore the frictional power, because the water required is in proportion to the temperature of the condensed steam produced, and therefore instead of feed-water at 100 degrees Fahr. and a vacuum, it will be better to have feed-water at 200 degrees Fahr. and no vacuum, and no air pump, and a reduced size of condenser and circulating pump.—Burgh.

CHAPTER B.

LOCOMOTIVE COMPOUND-ENGINES.

Webb's System consists of as Follows:—

The engine has two high pressure cylinders, attached to the outside frame plates between the middle and leading wheels, the connecting rods working on to crank pins in the trailing wheels, and one low pressure cylinder placed between the main frames at the front end of the engine, the connecting rod being attached to the single throw crank of the middle pair of wheels.

The steam is supplied from a regulator in the dome to a T pipe on the smoke-box tube plate, and thence by two copper pipes, down each side of the smoke-box through the back plate of the low pressure cylinder, and between the frames to the high pressure cylinders; the exhaust steam is returned by two pipes running parallel with the others into the smoke-box, and each pipe is carried round the inside of the smoke-box and enters the low pressure steam-chest on the opposite side; thus the pipes themselves are of sufficient capacity to act as a steam receiver, and being placed in the manner described, the exhaust steam is super-heated by the waste gas in the smoke-box, the final exhaust escaping on either side of the low pressure steam-chest, and thence into the chimney in the usual way, with this difference, that there are only half the number of blasts to urge the fire compared with the ordinary engines, and yet the engine steams very freely with the blast nozzle

Arrangement is also made so that steam direct from the boiler can be admitted to the low pressure cylinder for use when starting, but a relief valve is applied in connection with the steam chest, so that the pressure may never exceed about half that carried in the boiler.

London and North Western Railway.

Load of 10.0 a.m. Scotch Express, Euston to Carlisle.

	No.	Vehicle	ts.	ct.	qr.
Euston to Crewe.		Engine and Tender No. 300 when starting from Euston	62	13	0
	92	W. C. J. S. Van	11	2	3
	279	,, Third	11	7	0
	128	,, Single tri-compo.	12	10	3
	206	,, Coupé Compo...	12	7	3
	123	,, Day Saloon ..	12	4	2
	126	,, Single tri-compo.	12	10	3
	16	,, ..	10	11	2
	127	,, ..	12	10	3
	162	,, Van ..	11	18	0
	45	,, Single tri-compo.	10	9	2
	131	,, ..	12	10	3
	76	,, Van ..	11	2	3
	*43	,, Single tri-compo.	10	9	2
		Gross Load ..	214	9	1

* This Vehicle attached at Crewe.

	ts.	ct.	qr.
Load exclusive of engine and tender ..	151	16	1

This represents empty vehicles, the weights of Passengers and Luggage not having been taken.

Coal Consumed—79 cwts. = 29·46 lbs. per mile.
Less 1· 2 ,, for lighting up.

Coal actually consumed in running train .. 28·26 per mile.

Evaporation of Water.

Water used = 7546 gals. × 10 = 75,460 lbs.

$\frac{75{,}460 \text{ lbs. used.}}{8{,}848 \text{ ,, Coal.}}$ = 8.5 lbs. water evaporated per lb. coal.

The engine in question has two 13-inch cylinders and one 26-inch, the stroke in each case being the same—24 inches the driving wheels, 6 ft. 6 in. diameter, and the boiler pressure 160 lbs. From the time the compounds commenced regular working they had run in the aggregate 2,226,112 miles, the average consumption of coal, including that allowed for raising steam, being 29·1 lbs. per mile.—[WEBB.

WORSDELL'S SYSTEM CONSISTS OF AS FOLLOWS:—

Both cylinders are inside the frames, and are on the same centre lines as those of the ordinary eighteen-inch (18 in.) high-pressure cylinder engines. The left-hand cylinder is the high pressure one, being eighteen inches (18 in.) diameter, and the right-hand is the low pressure cylinder, twenty-six inches (26 in.) diameter, both being twenty-four inches (24 in.) stroke. The exhaust steam passes out of the high pressure cylinder through an en- copper pipe carried round the upper part of oke-box, so as to act as a super-heater for the steam before it enters the low pressure steam chest; it terminates in this low pressure steam chest, and the exhaust from the low pressure cylinder is passed out through the chimney in the ordinary way. A one-and-a-half-inch (1½ in.) starting valve is connected to this pipe just above the steam chest of the low pressure cylinder. This pipe conveys steam direct from the boiler, and is controlled by a small regulator on the driver's side of the engine, this having a spring handle so that it cannot be kept open, and a very slight opening is all that is necessary on this large piston to start the engine. To prevent this starting valve operating against the high pressure piston a special cut-off valve is arranged, which the driver pulls over, and which is thrown back by the exhaust steam from the high pressure cylinder, so that there is no attention required to this. This part of the arrangement is most successful. To prevent getting too high a pressure in the low pressure cylinder when steam is admitted from the boiler, an inch-and-a-quarter (1¼ in.) relief valve is fitted at each end, set to blow off at 80 lbs. pressure per square inch.

Great Eastern Railway.

Tabular Statement, giving principal dimensions of G.E.R. Compound Engine.

Cylinders—*High Pressure.*	ft.	ins.
Diameter of cylinder	1	6
Stroke of ,, ...	2	0
Length of ports..	0	11¾
Width of steam ports	0	1¾
,, exhaust ,, ...	0	3½
Distance apart of cylinders, centre to centre	2	0
Distance—centre line of cylinder to valve face	1	1
Distance—centres of valve spindles ..	2	0
Lap of slide valve	0	1⅛
Maximum travel of valve	0	5
Lead of slide valve	0	0 3/16
Low Pressure.		
Diameter of Cylinder	2	2
Stroke of ,, ...	2	0
Length of ports..	1	2
Width of steam ports	0	2
,, exhaust ,, ...	0	3½
Centre line of cylinder to valve face ..	1	5⅜
Lap of slide valve	0	1⅛
Maximum travel of valve	0	5
Lead of slide valve	0	0 3/16

VALVE MOTION—*Joy's Patent.*

Diameter of piston rod	0	3
Length of slide blocks	1	3
,, connecting rod between centres	6	10
,, radius rod	4	9½

WHEELS AND AXLES.

Diameter of driving wheel	7	0
,, trailing ,, ...	7	0
,, bogie ,, ...	3	1
Distance from centre of bogie to driving	10	9
Centres of bogie wheels	6	3
,, driving to trailing	8	9
Distance from driving to front of fire-box	2	0
Distance from centre of bogie to front of buffer plate	4	8½
Distance from trailing to back buffer plate	4	3

CRANK AXLE—Steel.

Diameter at wheel seat	0	9
,, bearings	0	7½
,, the centre	0	7
Distance between centres of bearings	3	10
Length of wheel seat	0	8
,, bearing	0	9

TRAILING AXLE—Steel.

Diameter at wheel seat	0	9
„ bearings	0	7½
„ centre	0	7
Length of wheel seat	0	8
„ bearings	0	9
Diameter of outside coupling pins ..	0	4¼
Length „ „ „ ..	0	4
Throw „ ..	0	12

BOGIE AXLE Steel.

Diameter at wheel seat	0	7½
„ bearings	0	6
„ centre	0	5
Length at wheel seat	0	7
„ bearing	0	9
Centre to centre of bearings	3	7
Thickness of all tyres on tread ..	0	3
Width „ „ „	0	5⅝

FRAMES—Steel.

Distance apart of main frames ..	4	0
Thickness of frame	0	1
Distance apart of bogie frames ..	2	7½
Thickness „ „ „	0	0¾

	ft.	in.
Boiler.		
Centre of boiler from rails	7	6
Length of barrel	11	$5\frac{1}{4}$
Diameter of boiler outside	4	2
Thickness of plates (steel)	0	$0\frac{7}{16}$
„ „ smoke-box tube plate ..	0	$0\frac{3}{4}$
Lap of plates	0	$2\frac{1}{2}$
Pitch of rivets	0	$1\frac{13}{16}$
Diameter of rivets	0	$0\frac{13}{16}$
Fire-Box Shell—Steel. Length outside	6	0
Breadth „ at bottom	3	11
Depth below centre line of boiler ..	5	6
Thickness of front plates	0	$0\frac{1}{2}$
„ „ back „	0	$0\frac{1}{2}$
„ „ side „	0	$0\frac{1}{2}$
Distance of copper stays apart ..	0	4
Diameter „ „ „	0	1
Inside Fire-Box—Copper.		
Length at the bottom inside	5	4
Breadth „ „ „	3	3
Top of box to inside of shell	1	4
Depth „ inside	6	$2\frac{1}{2}$
Tubes.		
Number of Tubes, 201		
Length „ „	11	$9\frac{1}{4}$
Diameter outside	0	$1\frac{3}{4}$
Thickness, No. 11 and No. 13 W.G.		
Diameter of exhaust nozzle	0	$5\frac{3}{4}$
Height from top of top row of tubes..	0	2
„ of chimney from rail	12	11

Heating Surface.

	sq. ft.	ins.
Of tubes	1082	5
„ fire-box	117	5
Total ..	1200	0
Grate area	17	3

Weight of Engine in Working Order.

	tons.	cwts.	qrs.
Bogie wheels	14	15	2
Driving „	14	16	2
Trailing „	14	18	0
Total ..	44	10	0

Weight of Engine Empty.

Bogie wheels	13	7	0
Driving „	14	3	2
Trailing „	13	12	1
Total ..	41	2	3

The Tender holds 5 tons of coal and 2755 gallons of water.

Trial No. 1.

	Mean effective pressure in lbs. per sq. inch.
High-pressure cylinder, front end ..	88·9
,, ,, ,, back ,, ..	83·1
Low-pressure ,, front ,, ..	37·0
,, ,, ,, back ,, ..	42·5

Full gear.

Indicated horse-power, high-pressure	237·63
,, ,, ,, low-pressure	225·10
Total ..	462·73

Speed 22 miles per hour. Load 6 carriages.

Trial No. 2.

	Mean effective pressure in lbs. per sq. inch.
High-pressure cylinder, front end ..	44·1
,, ,, ,, back ,, ..	43·4
Low-pressure ,, front ,, ..	22·6
,, ,, ,, back ,, ..	26·3

Cut off 62 *per cent.*

Indicated horse-power, high-pressure	256·51
,, ,, ,, low-pressure	264·33
Total ..	520·84

Speed 42 miles per hour. Load 12 carriages.

Trial No. 3.

			Mean effective pressure in lbs. per sq. inch.
High-pressure cylinder,	front end	..	38·5
,, ,, ,,	back ,,	..	36·4
Low-pressure ,,	front ,,	..	15·8
,, ,, ,,	back ,,	..	20·9

Cut off 50 *per cent.*

Indicated horse-power, high-pressure	294·90
,, ,, ,, low-pressure	304·84
Total ..	599·74

Speed 63 miles per hour. Load 12 carriages.

[Worsdell.

G. E. R. LOCO'. DEPARTMENT, STRATFORD.

November 22nd, 1886.

Comparative Statement of Water evaporated by "Compound" Engine, No. 704, and "Ordinary" Engine, No. 565, when working the 9·3 a.m. Passenger Train, *ex* Liverpool Street to Norwich,

	October 4th.			October 18th.		
	Compound Engine, No. 704.			Ordinary Engine, No. 565		
	Cwt.	qr.	lbs.	Cwt.	qr.	lbs.
Coal consumed	24	3	8	30	3	0
" " lbs. per mile ..	24·3			30·2		
Total quantity of water evaporated in Gallons	2196			2853		
Quantity of water evaporated per lb. of coal	7·9 lbs.			8·2 lbs.		
Average pressure of steam per square inch	138 lbs.			122 lbs.		
Temperature of feed water ..	64° F.			65° F.		
Average quantity of water injected per 5 minutes	112·5 galls.			126·4 galls.		
Load from London to Ipswich ..	14			15		
" " Ipswich to Norwich ..	6			7		

Remarks:—

On 4th inst., the Engine steamed freely, weather very favorable.

On 18th inst., the Engine steamed moderately, weather rather unfavorable.

S. WORSDELL, G.N.R. and JAS. HOLDEN, S.E.R.

CHAPTER C.

NATURAL AND INDUCED DRAUGHTS. (MARTIN.)

	Natural.	Induced.
Date of Trial	June 30th, 1886	July 5th, 1886
Duration of Trial	4 hours	4 hours
Lbs. of Coal consumed ..	350	500
,, ,, per square foot of Grate per hour ..	13	18·5
,, of Water evaporated, temp. of feed 72° ..	2788	5678
,, of Water evaporated per square foot of Grate per hour	103·2	210
,, of Water evaporated at 100° ditto	106	216
,, of Water evaporated at 212° ditto	118	240·75
,, of Water evaporated per lb. of Coal, temp. of feed 72°	7·96	11·35
,, of Water evaported per lb. of Coal, temp. of feed at 100°	8·19	11·65
,, of Water evaporated per lb. of Coal, temp. of feed at 212°	9·12	13
,, Pressure of Steam..	60 lbs.	70 lbs.
Cubic foot of Air per lb. of Coal	214	547
Cubic foot of air per square foot of Grate	47	169
Temperature in tubes 2 foot down at smoke box end	600°	750°
Temperature in Funnel ..	500°	450°
,, ,, Furnace ..	1300°	1950°
,, ,, Combustion Chamber	900°	1800°
Description of Coal	Nixon's Nav.	Nixon's Nav.

Single Flue Steel Marine Boiler—

Length	6′ 0″
Dia.	5′ 6″
Dia. of Flue..	2′ 3″
No. of Tubes	44
Dimensions	4′ 6″ × 2¼″ dia. inside
Area of Fire Grate	6·75 square foot

Heating Surface—

Tubes	97·3
Furnace	17·0
Combustion Chamber	38·0
	Total 152·3

Heating surface per square foot of grate ..	22·5
Diam. of Fan	2′ 0″
Size of Inlet	1′ 0″
Revolutions of Fan	1250
Revolutions of Fan Engine	250

The fire door consists of a curved plate that bends towards the front of the boiler. The size of the plate is that it can pass to and fro in the opening in the frame, so that the door has really a pendulous motion, and is retained in any position by counterbalances on the suspension rod. It will, therefore, be understood that the action of the use of this door is that when the door projects outwards from the mouth of the frame the air ascends to the top of the fire, and thus reduces the temperature of the same, and the speed of the draught is consequently reduced, but when the door is placed inwards or projecting inside the mouth of the frame, then the door acts as an inducer of draught, which, of course, is in proportion to the area of the opening. the door consist of a reduced draught by opening outwards, and an increased draught by opening inwards.

The fan is situated in the uptake of the smoke box, and is driven by a small engine; the gearing being frictional grooved, the smaller wheel is within the driving wheel, so that the rim is plain or smooth on the outside, and the grooves, therefore, inside the rim.—MARTIN.

NATURAL AND FORCED DRAUGHTS.

(SENNETT, INST. NAV. ARC'TS.)

	Natural Draught.	Forced Draughts.				
Steam pressure in boiler, lbs.	85·35	107·8	113·09	93·06	92·74	89·21
Mean air pressure in stoke holes, inches of water ..	—	2·02	1·52	1·4	1·89	0·55
Indicated horse power ..	5,588	6,628	3,370	11,158	9,544	11,722
Area of fire grate used in square feet	546	399	207	756	567	795
I.H.P. per square foot of fire grate	10·23	16·61	16·28	14·75	16·83	15·51
Heating surface per I. H. P. in square feet .. .. — Tubes	2·23	1·56	1·63	1·54	1·35	1·46
Heating surface per I. H. P. in square feet .. .. — Total	2·61	1·77	1·83	1·82	1·6	1·73
Coal used per I.H.P. per hour in lbs	2·39	2·48	2·6	2·2	..	2·16
Coal used per hour, in tons	5·96	7·33	3·92	11	..	11·30

CHAPTER D.

EVAPORATIVE POWER OF LIQUID FUEL.

(Admiral Selwyn.)

Diameter of Boiler	9′ 4″
Length of Boiler	9′ 8″
No. of Tubes in Boiler	120
Dimensions—Length	6′ 2″
,, Diameter, inside	0′ 3″
Diameter of Chimney	2′ 6″
Height from Uptake	20′ 0″
Diameter of Fire Tubes (2)	2′ 10″
Length of ,, ,, ...	6′ 2″

Cubical contents of Fire-Brick Furnace, about 6 cubic feet.

Area of Bottom of Furnace, corresponding to Grate Surface, each 3 square feet.

Chemical Constituents of Oil used, about carb. 86, hyd. 7·07.

Specific Gravity of Water being 1, from 1·050 to 1·080.

Weight of Oil in lbs. per cubic foot .. about 65 to 66 lbs.

,, Coal ,, ,, ,, 79 to 82 lbs. But cannot be stowed solid as this supposes.

Cubical contents in comparison with (Coal) ..	46 cubic feet.
Coal as Storage in Bunkers, per ton (Oil) ..	34 ,,
Amount of Oil used per hour in lbs.	230 lbs.
,, Water evaporated in lbs.	5040 ,,
Pressure of Steam in lbs. per square inch ..	55 ,,

Temperature of Feed Water .. No Feed during Experiment.

Diameter of Supply Oil Pipe, inside (2), $\frac{7}{32}$ of an inch, say 0·22.

Area of Steam Injector Pipe, inside (2), about 0·1429 of a square inch is the area open for steam, *i.e.* $\frac{1}{16}$ inch wide and 0·50 inches diameter outside the annulus.

Velocity of Air in Fire Box Door Opening, about the same as the steam.

Temperature in Fire Box	4000° F.
,, Smoke Box	520° F.

CHEMICAL COMPARISONS OF THE USE OF LIQUID FUEL COMPARED WITH COAL.

Liquid Fuel.

i.e. Oil alone, sp. gravity, 1050—1060.

Calorific value in lbs. of water vaporisable by one lb. of fuel.

When composition is			About
C 86 H 7·07	..	..	17 to 22·0 by theory.
the C V is 17·3.	..	..	12 to 14·9 in practice.

Coal.

About
15 to 16 by theory.
6 to 8 in practice.

Fluid Fuel.

i.e. including with the oil the hydrogen of the steam.

Pounds of water vaporisable by one lb. of oil and one lb. of steam.

About
23·4 to 29 by theory.
21·4 to 27 in practice. Loss of two units up funnel excluded.

Boiler.

Heating surface.

	Square feet.	
Furnace Tubes	66·0	Without any deduction for tubes or flat vertical surfaces, ash pits, &c.
Combustion Chamber	110·0	
Tubes	640·0	
Front Tube plate	15·0	
Total ..	831.0	

	Square feet.
Grate area	32·0
Tube	5·5
Funnel area	5·0

In the first case, where 21·4 is evaporated, from and at 212° F., 14·9 lbs. are due to the oil if it has a composition of

up fu) = 15·3 less again on eighth of the oxygen present to be deducted from the hydrogen, *i.e.* 0·8) 15·3 — 0·8 = 14·5 total

But as we have usefully burnt the hydrogen in the one lb. of steam employed with the oil as this hydrogen exists in the steam in the same proportion as in water (viz. Oxygen 89%
× 64=7·04 units of heat or lbs. of
added to the 14·5 units arising from
7·04=21·54, which is the evaporation
lbs. of oil and 230 lbs. of steam
this common 40 N. H. P. marine boiler)
evaporate by actual measurement 5,040 lbs. water per hour.

In this boiler it may be observed that 27 lbs. of oil per hour just suffices to keep the 55 lbs. pressure without blowing off at the valves. That with 57 lbs. of oil there is an evaporation of 100 lbs. of water per hour, therefore if the 27 lbs. oil per hour be regarded as just
in the latter case 57—27=30 lbs. of oil evaporating 1,000 lbs. of water, and this is partly accounted for by the fact that with this quantity of oil the temperature in the
if at all, higher than the temperature of th
say 280° F., thus there is no loss of heat up the funnel and about 2 units of heat (or lbs. water vaporisable per pound of oil) are economised. With a fuel of this nature the same economy might be produced where the full quantity of fuel is burnt by a greater number of smaller tubes giving more heating surfac
combustion in the case of injected fluid fuel, but only serves to take off the burnt gases and there is no ash dust or smoke with them.—Selwyn.

CHAPTER E.

STATISTICAL RECORDS OF MODERN MARINE COMPOUND ENGINES.

Name of Firm	John Penn & Sons.	
Name of Ship	" *Curlew.*" (Twin-Screw.)	
Consumption of fuel in lbs. per square foot of grate surface	44·9	With forced draught.
Consumption of fuel, per I.H.P. per hour	3·1	
Indicated H.P., total	1262	
Speed of Ship in knots, per hour ..	14·5	
Pressure of Steam in boiler	101	
Diameter of high pressure cylinder, No. 1	19	
,, low ,, ,, No. 2	37	
Length of stroke of piston, in inches ..	21	
Length of cut-off .. .. .. No. 1	15·5	
,, ,, ... No. 2	14·7	
Indicated pressure, *initial* .. No. 1	87	
,, ,, ... No. 2	11	
Indicated pressure, *exhaust* .. No. 1	51	
,, ,, ,, ... No. 2	0	
Indicated pressure, *mean* .. .. No. 1	54	
,, ,, ,, ... No. 2	13·3	
Speed of piston in feet, per minute ..	650	

Statistical Records of Modern Marine Compound Engines.

Name of Firm	John Penn & Sons.	
Name of Ship	"*Warspite.*" (Twin-Screw.)	
Displacement	7390 tons.	
Consumption of fuel in lbs. per square foot of grate surface ..	43½	With forced draught.
Consumption of fuel, per I.H.P. per hour	2·9	
Indicated H.P., total	10241	
Speed of Ship in knots, per hour ..	17¼ knots.	
Pressure of Steam in boiler	92½	
Diameter of high pressure cylinder, No. 1	55	
,, low ,, ,, No. 2	77	
,, low ,, ,, No. 3	77	
Length of stroke of piston, in inches ..	48	
Length of cut-off .. .. .. No. 1	33·6	
,, ,, ... No. 2	28·8	
,, ,, ... No. 3	28·8	
Indicated pressure, *initial* .. No. 1	34	
,, ,, ,, ... No. 2	10	
,, ,, ,, ... No. 3	11	
Indicated pressure, *exhaust* .. No. 1	45	
,, ,, ,, .. No. 2	2½	
,, ,, ,, .. No. 3	2½	
Indicated pressure, *mean* .. No. 1	45·9	
,, ,, ,, .. No. 2	13·1	
,, ,, ,, .. No. 3	14·1	
Speed of piston in feet, per minute ..	700	

These Engines are of the three Cylinder type—one high pressure, and two low pressure Cylinders in each set.

Statistical Records of Modern Marine Compound Engines.

Name of Firm	I & G. Rennie
Name of Ship	"*Swallow.*"
Indicated H. P., Total	1567
Speed of Ship, in knots, per hour ..	13·42
Pressure of Steam in Boiler ..	100 lbs.
Diameter of High Pressure Cylinder	22"
Diameter of Low Pressure Cylinder	42"
Length of Stroke of Piston, in inches	24"
Indicated Pressure, Mean No. 1 ..	56·02
Indicated Pressure, Mean No. 2 ..	13·78
Speed of Piston, in feet, per minute	639·8

Twin Screws, Two Steam Cylinders of each size.

Average air-pressure in boiler-rooms equal to 1·67 inches of water.

Statistical Records of Modern Marine Compound Engines.

Name of Firm	I. & G. Rennie
Name of Ship	"*Calypso.*"
Area, in square feet, of Midship Section	510
Consumption of Fuel, in lbs., per square foot of Grate Surface, per hour	28·9
Consumption of Fuel, per I. H. P., per hour	2·72
Indicated H. P., Total	3200·3
Speed of Ship, in knots, per hour ..	14·896
Pressure of Steam in Boiler ..	85·7 lbs.
Diameter of High Pressure Cylinder	42″
Diameter of Low Pressure Cylinder	72″
Length of Stroke of Piston, in inches	36″
Indicated Pressure, Mean No. 1 ..	35·7
Indicated Pressure, Mean No. 2 ..	13·655
Speed of Piston, in feet, per minute	502·74

Tandem Engines, Two Cylinders of each size.

Statistical Records of Modern Marine Compound Engines.

Name of Firm	I. & G. Rennie
Name of Ship	"*Canada.*"
Area, in square feet, of Midship Section	506
Consumption of Fuel, in lbs., per square foot of Grate Surface, per hour	27·68
Consumption of Fuel per I. H. P., per hour	2·88
Indicated H. P., Total	2429·5
Speed of Ship, in knots, per hour	14·17
Pressure of Steam in Boiler ..	58·37 lbs.
Diameter of High Pressure Cylinder	36″
Diameter of Low Pressure Cylinder	64″
Length of Stroke of Piston, in inches	30″
Indicated Pressure, Mean No. 1 ..	39·31
Indicated Pressure, Mean No. 2 ..	11·30
Speed of Piston, in feet, per minute	524·7

Tandem Engines, Two High and Two L. P. Cylinders.

(INST. NAV. ARCT'S.)

Name.	Engines.						Steam Pressure.	Builders.
	No. of Cylinder.	Diameters. Ins.				Stroke. Ins.		
African . .	3	21	34½	55½		36	150	T. Richardson & Sons
Alcides . .	3	29	43	68		54	150	J. & J. Thompson
Anglian . .	3	26	42	69		42	150	T. Richardson & Sons
Cleveland . .	3	21	35	57		39	150	Centr. Marine E. Co.
Clitus . . .	3	23½	37	61		42	150	Palmer's Co.
Condor . . .	3	11	16½	30		21		Cox & Co.
Dalhousie . .	3	10½	15	27		16		W. B. Thompson
Danube . .	3	15	23	38		24		D. J. Dunlop & Co.
Dunbrodie . .	3	24½	39	62		42	150	Lond. & Glasgow Co.
Eddystone . .	3	23	37	58		48	135	W. B. Thompson
Euterpe . .	3	24	42	69		48		T. Richardson & Sons
Fijian . . .	3	18½	31	49		36	150	Palmer's Co.
Gloamin . .	3	19	30	50		42	170	Hall, Russell & Co.
Hubbuck . .	3	27	42	69		43	150	T. Richardson & Sons
Inishowen Head .	3	24½	37	64		48	160	Harland & Wolff
Iran . . .	3	28	43	77		51	160	Harland & Wolff
Jumna . .	4	35	48½	67	94	60	160	Denny & Co.
Lahora . .	4	24	34	48	68	42	160	Denny & Co.
Libra . . .	3	25	42	67		42	150	Wallsend Slipway Co.
Loch Etive . .	3	21½	34	56		42	165	Gourlay, Bros. & Co.
Lusitania . .	3	36	60	96		48	150	T. Richardson & Sons
Mandalay . .	3	20	33	54		36	160	Blair & Co.
Mercedes . .	3	21	35	57		39	160	Blair & Co.
Monmouthshire .	3	30	47	70		51	140	London & Glasgow
Northenden . .	3	21½	35	57		39		Wallsend Slipway Co.
Orizaba . .	3	40	66	100		72		Barrow S. B. Co.
Paumben . .	3	18	30	48		36	160	A McMillan & Son
Port Pirie . .	3	29	44	74		48	150	Wingham Richardson
Powhatan . .	4	18	38	60		42		Barrow S. B. Co.
Racer . . .	3	15	23	40		27	150	Carr & Co.
Roseland . .	4	6	9	16		12		Cox & Co.
Saint Filans .	3	24½	37	64		48	160	Harland & Wolff
Saint Oswald .	3	24	39	64		42	150	Wallsend Slipway Co.
Satelite . .	6	10½	16	26		22	150	D. J. Dunlop & Co.
Scholar . .	3	20	33	54		36	160	Blair & Co.
Teresa . . .	3	18	30	48		36	155	A. McMillan & Son

Particulars.	" *Inflexible.*"			" *Colossus.*"		" *Phaeton.*"
Description of Engines	3 cylinder vertical compound.			3 cylinder vertical compound		Horizontal compound
Diameters of cylinders in ins. — High pressure	2 of 70"			2 of 58"		2 of 42"
Diameters of cylinders in ins. — Low pressure	4 of 90"			4 of 74"		2 of 78"
Length of stroke ft. ins.	4' 0"			3' 3"		4' 0"
Propeller — Description ..	2 bladed			4 bladed		4 bladed
Propeller — Diameter ft. ins.	20' 2¼"			17' 8½"		14' 0½"
Propeller — Pitch ft. ins.	23' 0½"			18' 7¼"		20' 1¾"
Boilers — Number	12			10		8
Boilers — Description	FOUR EACH OF Oval 3 furnace	Oval 2 furnace	Oval Double ended 4 furnace	EIGHT Oval 3 furnace	TWO Oval 2	Cylindrical high 3 furnace
Boilers — Transverse dimensions	13'7"×15'6"	11'1"×13'4"	9'4" × 14'3"	12'9"×15'3"	7'10"×14'0"	18' 5" dia.
Boilers — Length	9' 0"	9' 0"	17' 0"	9' 9"	9' 9"	9' 8"
Load on safety valves lbs.	60			64		90
Furnaces — Number	36			28		24
Furnaces — Diameter	twelve of 3' 6"	eight of 3' 3"	sixteen of 3' 6"	Twenty-four 3' 5"	Four of 2' 10"	18 of 3' 3" 6 of 3' 0"
Furnaces — Length	6' 0"	6' 0"	6' 6"	6' 9"	6' 9"	7' 0"
Furnaces — Grate area in sq. ft.	829			645		546
Heating surface of boilers in sq. ft. — Tubes	18,654			14,745		12,456
Heating surface of boilers in sq. ft. — Total	22,288			17,507		14,562
Area through tubes in sq. ft. ...	158			117		87·5
Funnels — Number	2			1		2
Funnels — Size	Oval 10' 0" × 8' 0"			Oval 12' 0" × 8' 0"		8' 0" dia.
Funnels — Height above fire bars	70' 3"			1		61' 8"
Ratios of — Tube heating surface / Grate area	22·5			22·8		23 3
Ratios of — Area through tubes / Grate area	·190			·181		·160

TABLES, 1886.—INST. NAV. ARCT'S.

"Mersey."	"Scout."	"Rodney" and "Howe."	"Trafalgar" proposed
Horiz.ntal compound	Horizontal compound	3 cylinder vertical compound	Vertical triple expansion.
2 of 38"	2 of 26"	2 of 52"	2 of 43" 2 intermediate of 62"
3 of 64"	2 of 46"	4 of 74"	2 of 96"
3' 3"	2' 6"	3' 9"	4' 3"
3 bladed	3 bladed	4 bladed	not yet decided
13' 0"	10' 6"	15' 6"	
18' 5⅜"	12' 6"	19' 6"	
6	4	12	6
Low Cylindrical 3 furnace	Low Cylindrical 3 furnace	Oval 3-furnace	High Cylindrical 4 furnace
10' 0" dia.	9' 3" dia.	11'0" × 15'0"	16' 2" dia.
18' 9"	17' 10"	9' 8"	10' 3"
110	120	90	135
18	12	36	24
3' 2"	2' 10"	3' 0"	3' 7½"
7' 0"	6' 0"	7' 0"	7' 4"
399	207	756	609
10,367	5,500	17,174	17,040
11,700	6,170	20,294	19,390
61	32	102	96
1	1	2	2
7' 2" dia.	6' 6" × 4' 9"	9' 0" × 5' 6"	7' 0" dia.
52' 6"	55' 0"	75' 0"	65' 0"
25·9	26 5	22·7	28
·152	·154	·134	·158
·100	·125	·114	·126
4	4	8	6
5' 0"	3' 6"	5' 0"	5' 6"

WYLLIE'S NOTES ON THE TRIPLE COMPOUND MARINE ENGINE.

INST. MECH. ENGR'S., 1886.

The most important conditions to be considered in order o obtain an efficient engine, are that there should be approxi-nate equality; firstly, in the range of temperature in each ylinder; secondly, in the initial stress on each crank; and hirdly, in the indicated horse-power of each engine. What ay be termed the complements to these three essentials are :—(1), steam jacketed cylinders; (2), cylinder ratios; (3), velocities of initial and exhaust steam; (4), clearance and compression; (5), receiver capacity; (6), piston speed; (7), order of sequence of cranks. Marine engines should be so designed that any working part could be easily examined or removed; and this is impossible with a tandem engine. The arrangement of cylinders on three cranks fulfils the required conditions more nearly than any other design.

The comparative economical results obtained from the working of three steamers with triple expansion engines were set out as follows:—

Type of Engines.	Compound.	Compound.	Triple.
Boiler pressure lbs. per sq. in.	70	76	140
Speed, knots per hour ..	9	9½	10
Indicated h.p., total	660	790	890
Coal, consumption per day, tons	15½	18	13½
Ditto, per 1 h.p., per hour, lbs.	2·19	2·13	1·41
Quality of coal used—	German.	Cardiff.	Mixed.

CHAPTER F.

ELECTRICAL POWER.

The Amount of Indicated Horse-power to furnish sufficent Electrical Energy to light Edison-Swan's Twenty Candle-power Lamps.

Dynamo Machines are generally built to give out, at a definite speed, a definite quantity of Electrical Energy. This quantity may be doubled, trebled, &c., according to the size of the Machine, and is called a unit, consisting of 1,000 Watts (a Watt is the electrical unit of power and is current ex-electromotive force.)

Machines are therefore classified as "One Unit Machine." Two Unit Machine," &c., the purchaser instantly knows what is the capabilities of the Machine, at its fixed speed, and can easily calculate the number of Twenty Candle-power Edison-Swan Lamps it will light.

For example, a "Three Unit Machine" will give out 3,000 Watts and Edison-Swan Lamps take Three Watts per Candle-power, therefore a Machine of this size will give out 1,000 Candle-power, which is equal to 50 Twenty Candle-Power Lamps, or 100 Ten Candle-power Lamps.

Now one Indicated Horse-power equals 746 Electrical Watts; therefore 3,000 Watts require 4·02 I. H. P., this of course does not take into consideration the friction due to bearings, and brushes upon the commutator, but including friction about 4·25 or 4¼ I. H. P. will be sufficient power to run a "Three Unit Machine" at its normal speed with a full load of 3,000 Watts, that is 50 Twenty Candle-power Edison-Swan Lamps.—SWETE & MAIN.

PREFACE.

THAT the need of a practical work on "Compound-Engines," has become so real in its requisition, is the reason why this "Pocket Book" has been written; in fact, so much has the want been apparent, that, the query is, "why has it not been done before?" and, if the leading manufacturing firms had not put their experience so fully into my hands, I could not have done it now.

I have dealt with the subject as largely as convenient, and at the same time, have not, I believe, omitted any portion demanding attention.

The formulæ introduced are for the main portion new, but at the same time founded on results absent from dispute, so that the future engine can be safely designed from them, without doubt of the present results and it may be with improved economy.

I recommend that more care be taken in the proportions of the steam ports and their valves, and more especially in the expansion gear, that it takes into consideration the *time* in connection with the *motion* of the piston.

I recommend also, that young engineers should

"fathom" the depths of the "Science of Steam," so that they may understand the article they are using, and not as they now too often do, "work in the dark," by "doing as others have done," and resting with content thereon.

I firmly believe that the period is close at hand, when the "steam" as used at present, will be known as a fallacy, and that a new element from the same source will be generated. In connection with this, I am certain that the late Professor Rankine saw it, when he introduced the term "steam gas."

What is wanted to drive an engine—as the phrase is—is a nearly frictionless and self-containing power; not as now is, a self-consuming vapour, that leaves behind it all the evils possible for the next volume to make good.

The fact is, that in due course of time, and that not long, engineers will "wake up" from their lethargy, and consider what heat really is, and thus understand, that Electricity is life to themselves, and therefore to all uses and purposes at their command.

I will not blame my profession, because much change has not taken place lately, knowing as I do, that commercial matters to a great extent regulate progress, but, I trust, the time is not far off, when a struggle will be made by some one, to save that heat which is now lost by malformation, and waste afterwards.

What the future engine must be, is a machine capable of containing the electricity in the steam during its development of power, and the only way of aiming at that state of improvement is to first make the steam properly, and secondly, use it in the same manner afterwards.

It will be found that this is explained in this work, and I trust so as to be understood, and of equivalent value.

The formulæ for the unit of heat is my own, and is I believe reliable, from the fact, that the weight of the steam used is multiplied by the temperature, which of course embraces the constituents of the volume.

The steam constant I have introduced is new also, and has been carefully worked out from the best modern examples of compound-engines, as will be seen from the calculations in both cases.

The tables of scientific results are of much value, and the figures given may be taken as standards for further practice and improvement.

Observing the manufacture and material of future engines, I am aware of the destructiveness of high-pressure steam—say at 200 and 300 lbs. per square inch—on the material now used, but on the face of that, from experience, I am certain that wrought iron and other mallable metal, can be successfully applied when the ingredients are properly mixed and manufactured. As for example, a "fine" wrought iron cylinder or valve facing, will "stand" when the hardest cast iron material will surfarise and create metallic plumbago, all from the fact of the disturbance of the "layers" of the planed or bored metal.

What is really requisite is, that the temperature of the finished surface should be as the steam is, that is to come in contact with it hereafter, and not as now is "finish cold and use it hot" afterwards. A little thought of contraction and expansion properties will assist to appreciate this.

I, of course, could enlarge on those and other matters, but as space is paramount now, I conclude with the belief that this work is the most valuable I have written, and may be considered as a prelude to my larger work "Science of Steam."

N. P. BURGH.

LONDON, 80, CORNHILL, E.C.,
October, 1876.

CONTENTS.

RICHARD TILLING,
STEAM PRINTER & STATIONER,
WARNER STREET, SOUTHWARK,
LONDON.

BURGH'S POCKET BOOK

OF

COMPOUND ENGINES.

CHAPTER I.

WHAT IS STEAM?

IT has often been expressed by some scientific engineers that as a practical result of improvement over the ordinary expansive engine, the compound-engine is a myth. This delusion, it is now our purpose to dispel, and to clearly show what the compound-engine is at present, and what, so far as we know now, it will be in future.

Steam being the motive power in the cylinders of the compound-engine, it is obvious we must first explain what is steam.

Taking this subject as steam proper, we should be contented by stating steam is heat, water and air, but, unfortunately, steam is not made "proper," so we have, therefore, to deal with what is produced as a total.

This "total" consists of heat, water, air to equal steam, then steam produces pressure from which temperature is realised, producing thereby radiation, which is another name for cooling, causing liquefaction or saturation in the cylinders, often termed condensation. This condensation gives out friction, that affects the piston speed, and reduces the indicated horse-power.

Next, what are the now known remedies for the objectionable parts of the "total" we have explained? They are steam jacketing and lagging, which is said to give out the results, to a great extent, of non-radiation, causing better "full" steam, more expansion, and a cleaner exhaustion.

Going back again to the query, "What is Steam?" we have now to explain further its properties.

Steam is an elastic gas of more or less density, according to the proportions of its constituents—as, for example, should the heat be reduced by cooling, the same quantity of water is increased *in its effect*, and thus the elasticity is reduced, the pressure of water in steam being in all cases a known constant, while the presence of heat is, more or less, increased or reduced by circumstances.

We may explain, further, that, given a pressure

of steam, and given a certain bulk (in the engine cylinders, mind), a certain amount of elasticity is adherent to those facts; but the moment the heat escapes the water remains and the elasticity is reduced thereby.

Again: Let it be supposed that the heat is increased in the steam before it enters the cylinder by extreme superheating—which, by the way, fifteen years ago was considered the *acme* of making steam—what will be the result? Why, the water will be rendered so infinitesimal that the elasticity will be reduced by that fact.

Steam, therefore, is a gas that should be made properly and not by guess work; and it is equally obvious that if the loss of heat destroys the perfection, the extreme increase of heat will do the same.

It is apparent, therefore, that steam should be elastic, because, if not so, a great loss of power occurs; as, for example:

Suppose low-pressure steam, say 30lbs. on the square inch, is initially used in a cylinder for half the piston's stroke, and expansion occurs up to seven-eighths of the stroke, we should exhaust the steam at about half the pressure it was at the point of cut-off, *theoretically*. Whereas, we find in practice that at least two-thirds of the heat in the steam is lost by radiation, or by expending itself on the internal surface of the cylinder, and, consequently, for at least half the stroke the original elasticity of the steam is lost.

Now, when the steam is not properly elastic, a film of water congregates on the internal surface of the cylinder, and friction of steam results in the reduction of the power.

We come now to the fact not only that steam has friction under all circumstances, but also to another, that there is a limit to that friction, and our purpose is to find it out and explain this.

We will not now introduce formulæ, in its proper place we shall do so.

The "limit of the friction" is entirely dependent on the elastic force of the steam, and the "elastic force" is dependent on the *proper* amount of heat in the steam in proportion to the water, and also in proportion to the air.

The "air," let it be remembered, has generally been overlooked in practice as a function in making steam; and considered, therefore, as an accidental property, of little, if of any, importance.

Truly, there have been a few inventors that have experimented on the effect of introducing air into the boiler as an auxiliary, but to the present nothing we know of has been done to ascertain the proper quantity in proportion to the heat and water.

What we mean by "the limit of the friction" is the least amount of power that is absorbed by the rubbing contact of the steam on the surface of the cylinder.

Next, we have to deal with "the elastic force," which is another name for "expansion." Or perhaps we shall make the matter clearer by explaining what "elastic" means: not the literal meaning, but the science of the term in relation to steam.

The "elasticity" of steam is the power it has *in itself* to enlarge in bulk, without losing its relative constitutional proportions.

Suppose, for example, heat is 100, water 20, and air 5, then the steam equals 125 in constitutional bulk.

Next, suppose the initial quantity is 500 and elasticises into 15,000, or 1 to 3, then remember the water is the same, the heat is the same, and the air is the same, supposing nothing be lost.

But should the steam, during its exertion to elasticise, change its proportions of constituency, then the elastic force is imperfect, and the liquefaction results.

What we mean here by liquefaction is that the

water and air amalgamate and form more water, until the heat 100 becomes 90, and the water 25.

We are, of course, treating now of the fact that no heat escapes, but rather that the change of proportionate constituency is due to the malformation of the steam.

What, then, must be the result of radiation as a comparison to liquefaction alone? Or, perhaps more conclusively, what must be the result of

We next come to the fact that if steam can lose its elastic force from imperfect proportions when theoretically made, how much more of that force is lost when the steam is practically made.

In answer to that, we speak from the best practice that, in the present year 1876, not more than *one-third* of the heat in the steam drives our best Compound Marine Engines.

We have, it will have been noticed, treated especially that "heat" is the main constituent in steam. It is pretty well known what air and water are, but not so as to heat. We have, therefore, to discuss, What is heat?

It is very evident that sensible heat is a resultant from "latent" heat, and the term latent is a happy conventional term used by philosophers, because it means "hidden."

As far back as the year 1762 Dr. Black is said to have discovered latent heat, in a practical form, as the origin of sensible heat, but he omitted to state what latent heat was, as also have his troop of disciples.

Now, then, what is "latent heat?" To answer that, let us go back over the ground again.

Sensible heat proceeds from latent heat, being therefore a resultant.

Next, as a resultant must come from a source there must be a supply, and that supply must have a third source beyond latency, because, if not, the sensibility would terminate.

Let us think, now, what keeps this globe of ours in motion? Why, light! as, for example, see Crooke's "Radiometer."

Then comes the question, Where does the light come from? Why, from electricity, *which is the gift of the Great Creator.*

Then, as light is a resultant from electrical action (which we know from the fact that intense lights emit intense temperature), we are therefore certain in concluding that heat is electricity in a slightly developed form. We have, therefore, the fact that the heat in steam is a germ of electricity; and that the more developed that germ is, the more elastic the force will be in the steam.

A Mr. Rowell, at Oxford, in treating of the electrical action in steam, states:—

"To me the only explanation appears to be, that the excessively heated particles of water attract electricity to such a degree, that each particle is so completely enveloped in its electric coating, that the particles do not actually come into contact with any body on which the steam impinges. If cooled or condensed, it ceases to be superheated steam.

"The expansive force of steam seems explicable in accordance with this view also, and no other. There is no proof that the particles of water actually expand from heat, and there certainly is no chemical change on the conversion of water into steam; but the intensity with which electricity is attracted by the surface of the superheated particles of water, thus forcing them apart, seems to be a sufficient cause for the expansive force of steam. It may be impossible to fairly estimate the force with which electricity is attracted to the surface of heated particles of water, but that it is great is shown by the fact that the most intensely heated water, when converted into steam, is cool and dry even when driven with enormous force on any object."

But we shall go farther even than that, by stating that the electrical action in steam, is partly due to the condition of the feed water. That is the power of the heat when generated into bad feed water, is, to a very great extent absorbed by the impurities, consequently the steam is surcharged with the result of that absorption.

It is very obvious, therefore, that steam can be made bad as well as good, or perhaps we may say weak steam and strong steam—the pressures being alike.

The preparation of feed water has, during the last three years, received a certain amount of attention; but we have found very great difficulty in getting the true value of the matter to be appreciated.

In a similar way also we have found it a very hard task to get it to be understood, that the quantity of water in a boiler, in proportion to the thermal value of the heating surfaces, to a very great extent affects the power of the steam.

As space here is of the utmost importance we have thus put our scientific conclusions in as brief a form as possible; but, we may add, those conclusions are the result of much experience and study.

CHAPTER II.

THE ACTION OF STEAM IN THE CYLINDERS OF COMPOUND-ENGINES.

THE action of steam in the cylinder of a steam engine has too often been understood as a matter of mere expansion—*i. e.*, length of cut-off—proportionately decides the amount of expansion, and thus the matter ends. There are, however, many important facts in operation, which, although generally overlooked, are there, or in being, and it is our purpose now to point these out.

Starting, then, we commence with initial steam, which is the scientific term for the new steam that enters the cylinder from the time of "lead" to the "cut-off." This bulk of steam, it must be remembered, is the total indicated power of the whole of the stroke of the piston in the small cylinder and the whole of the stroke of the larger cylinder—if only two cylinders are used; but if a third, then the power of the same steam extends to a third stroke of the piston.

Now, were it only the stroke of the piston—or, rather, the motion, which is direct—the steam had to deal with to develope the indicated power, the matter would be the simplest possible. But there is a second and a third motion to intervene before the power is developed.

The first motion, then, is direct for the piston. The second motion is vibratory for the connecting rod, and the third motion is rotary, or revolving for the crank pin.

The initial steam commences with the first motion, and the elastic force operates with the main portion of the second and about three-fourths of the third motion.

The travel of the crank-pin in relation to the vibratory action of the connecting-rod *governs* the development of the initial steam inversely.

That is, the steam, of course, moves the piston, and the piston moves the connecting-rod, and equally, of course, the connecting-rod moves the crank-pin; but, for all that, the motion of the crank-pin governs or affects the length proportion of its connecting details, and, therefore, the action of the steam is subservient to that fact.

It may now be said what we mean is that the proportion of the length of the connecting-rod to the length of the crank or half-stroke of the piston governs the action of the steam in the cylinders; but it is not alone that, it is more.

We must now direct attention to the fact that we are not dealing with one crank-pin but rather with two at the least, and on occasions with a third pin.

Our purpose, then, is to point out what the two pins are doing when the *same* steam is driving or moving them. But first we must dismiss the commencement of the action of the steam, or, rather, we must consider the steam is in full operation, and the engine in continuous motion, for this reason :—

When the first steam enters the cylinder and "starts" the engine, the larger cylinder is empty, and therefore its crank-pin is moved by the new steam instead of by the expansion of the initial steam.

Our reason for explaining such a simple fact as

young engineers

this subject, and are too apt to say "when the high-pressure piston begins to move," &c.

Suppose, next, that the engines are in "full run." The initial steam on entering the cylinder will impel the piston direct on to the crank-pin; but the centrifugal force and the momentum will cause the crank-pin to move either way, as the case may be.

The moment the connecting-rod leaves the central line of motion, then the initial steam lessens its direct effect on the crank-pin, because the pin absorbs a proportionate amount of frictional power due to the angle of the connecting-rod.

Or, perhaps, we shall make this clearer by explaining that initial steam at the end of the stroke of the piston causes a *blow* on to the crank-pin.

Now, then, it is just this "blow" that we want to mitigate, and for the reason that we cannot "get rid of it," and therefore resort to the best means of causing an equal pressure on a series of crank-pins for the entire revolution.

In fact, the whole affair of the steam-engine, to make it an economical machine, is to make it smooth working, and to do that we must understand our subject.

We left the crank-pin on the "rise," let us say, and the piston "moving on." The steam, let it be supposed, is cut off when the crank-pin has reached one-fourth of the half circumference. The elastic force here comes into operation, while, at the same time, the crank-pin is reducing its absorbing fractional powers, until, say, the vertical line is reached. Supposing this to be horizontal engines under explanation.

The crank-pin descends and the pin-friction

again increases—as it must when the crank ascends on the opposite side—until the "dead" centre is reached, or the horizontal line, termed, in this case, the "central" line of motion.

The expanding steam has by this time partially exhausted into a receiver of the valve casing of the low-pressure cylinder.

Now, then, comes the reason why the engine must be supposed to be in motion. When the crank-pin of the high-pressure cylinder is on the central line of motion, or is horizontal, the crank-pin of the low-pressure cylinder is vertical, or on the top or bottom of the circle, and the piston at half stroke, less the verved sine of the arc of position due to the length of the connecting-rod.

But we will now suppose the engines are in motion and "running" fairly—in fact, doing their average duty at an average speed.

A section of the two cylinders, ports, and slide-valves must now be imagined, and strict attention must be paid to the relative positions of the two pistons, at certain points, and also to the direction of the motions of the pistons, either when moving together or in opposite directions, and next must be "seen" the movement of the valves of each cylinder, in strict relation to each other.

When the high-pressure piston is at the end of its stroke the "lead" of the valve admits the steam into the cylinder. The valve and piston then move in the same direction for a certain length. The valve next has a reverse motion, by which means the supply-ports are closed; the expansion of steam then occurs until the exhaust side is open, when the steam exhausts into the low-pressure cylinder--the piston of which is moving in the same direction as the high-pressure piston—that is, the low-pressure piston is moving *fiom* the nearest end of its cylinder, and the high-pressure piston is moving *towards* the nearest end of its cylinder. The result is, that when the high-pressure piston has reached the end of its stroke the exhaust-ports of the high-pressure cylinder are full open.

When the commences to go

occurs—that is, the low-pressure piston continues its travel in the same direction as before, thereby allowing more room for the expansion of the steam until its slide-valve closes its supply steam-port.

It may here be observed that although the supp

ssure cylinder still remain open; the steam, therefore, that is in the high-pressure cylinder becomes compressed by its piston, and, therefore, escapes into the receiver, or low-pressure valve casing.

The low-pressure piston is now nearly in a line with the high-pressure piston, but the latter is commencing its new stroke, while the former is completing its old stroke.

When this stroke of the low-pressure piston is complete, its valve gives the lead, and the steam that was left in the receiver, or valve casing, and also the steam that was left in the high-pressure cylinder becomes a combined motive power for the low-pressure piston, as it commences its stroke, and this said steam continues its duty *alone*, until the exhaust-port of the high-pressure cylinder is sufficiently open to allow the new expanded steam

therefore, further impel the low-pressure piston.

This, then, is a brief explanation of the mechanical movements of the two pistons and their valves: we say "their valves," because the valve governs the steam that actuates the piston.

We will now explain the *relative* motion of the two pistons and valves, driving one and a-half revolutions of the crank-pin; it must be "one and a-half," because the cranks are at right angles.

Comparative Table of the Relative Motions of the Pistons and Valves of a Compound Engine.

High Pressure Cylinder.		Low Pressure Cylinder.	
PISTON.	VALVE.	PISTON.	VALVE.
At the beginning of its stroke.	Lead.	Three-eights of stroke going back.	Supply port full open.
Moving forward at three-eights of stroke.	Moving forward.	Moving back.	Moving forward.
Moving forward.	Full open.	At end of stroke.	Exhaust ports slightly open, supply port lead.
Going forward.	Going back.	Moving forward.	Moving forward.
Moving forward.	Cut off.	Moving forward one-tenth of stroke.	Supply and exhaust ports nearly open.
Moving forward at seven-eights of stroke.	Exhaust port about to open.	Moving forward one-fourth of stroke.	Supply and exhaust ports full open, valve still.
Moving forward to end of stroke.	Exhaust ports full open.	Moving forward.	Moving back.
Moving back.	Moving back.	Moving forward.	Moving back.
Moving back.	Moving back, exhaust port open.	Moving forward and at end of stroke.	Lead exhaust port opening.

Our next explanation relates to the action of the steam as the indicator diagrams are formed.

The first matter for consideration to be noticed in this case is that we must suppose that there are two indicators, one attached to each cylinder at the back and front ends. The high-pressure piston, when at the end of the stroke, receives the impact of the new steam, as likewise also does the piston of the indicator, and is therefore driven up accordant to the resistance of the spring above it.

The low pressure piston has performed nearly half of its stroke, and the expanding steam line of the indicator diagram is formed for that distance also.

The initial line of the high pressure indicator diagram, is now being formed, say for half the stroke when expansion commences.

Just before the cut-off in the high-pressure cylinder the steam in the low-pressure cylinder commenced second expansion. The piston of that indicator then began to gradually descend.

The low-pressure piston is moving back, and also loses the impact of the steam, because exhaustion commences, while at the same time expansion occurs in the high-pressure cylinder, and the indicator piston commences to descend also.

The high-pressure piston is now at nearly the

end of its stroke, but the length of the indicator figure is complete.

The length of the indicator figure was complete, also, in the low-pressure cylinder, when its valve commenced exhaustion, which was when the cut-off occurred in the high-pressure cylinder, therefore the length of the low-pressure indicator diagram is completed when the expansion line of the high-pressure diagram is being formed.

The exhaust port of the high-pressure cylinder is now being opened, the piston moves to the end of the stroke, while the indicator piston descends; complete exhaustion now commences; the indicating piston is supposed to be at rest, as also is the indicating piston of the low-pressure cylinder and the exhaust line of that diagram being half made.

The remainder of the exhaust line is finishing while the exhaust line of the high-pressure cylinder is commencing, and continues its completion until the piston has reached the end of its stroke.

The following table of the formation of the two diagrams, in combination with preceding description will be found of practical use:—

COMPARATIVE TABLE OF THE FORMATION OF HIGH AND LOW PRESSURE INDICATOR DIAGRAMS OF A COMPOUND ENGINE WITH THE TWO CYLINDERS SIDE BY SIDE.

HIGH PRESSURE DIAGRAM.	LOW PRESSURE DIAGRAM.
1 Not commenced.	1 Vertical end steam line described.
2 Not commenced.	2 Expanding horizontal steam line one-third formed.
3 Vertical initial steam line described.	3 Expanding steam line being continued.
4 Initial horizontal steam line commencing.	4 Second expansion commenced.
5 Initial horizontal steam line complete.	5 Exhaust line commenced vertical end, exhaust line described.
6 Expansion line commenced.	6 Exhaust line forming.
7 Expansion line ceased.	7 Exhaust line forming.
8 Exhaust line commenced vertical end, exhaust line described.	8 Exhaust line forming.
9 Exhaust line forming.	9 Exhaust line complete.
10 Exhaust line forming.	10 Compression or lead line formed.
11 Exhaust line forming.	11 Completed.
12 Exhaust line complete.	12 Completed.
13 Compression or lead line commenced.	13 Completed.
14 Compression or lead line completed.	14 Completed.
15 Completed.	15 Completed.

We have next to consider what the steam *suffers* from the *cooling* of the cylinders, because it must be remembered that if the steam expands, its temperature is due to the pressure in pounds per square inch.

As, for example :—Suppose steam at a pressure of 80lbs. on the square inch enters the cylinder and is cut off at one-third of the stroke.

At this point the temperature should be 322 deg. Fahr., which will, of course, be the heat of the internal surface of the cylinder.

At the point of exhaustion the steam will have expanded nearly three times, then the pressure will be about 26lbs. on the square inch at the most, and its temperature, 268 deg. Fahr., showing a loss of 54 deg. in the temperature of the cylinder.

Then comes the fact that if the cylinder is that much cooler, the new or succeeding initial steam must make good that loss, and is therefore *robbed* in so doing.

Or, in other words, the higher the expansion in a single cylinder the lower the surface temperature must be, for the new or next initial steam to "heat" or "make up" for the preceding loss.

These facts are, therefore, the cause why high-pressure steam is adherent in compound-engines.

Also is the fact that the higher the initial

pressure and the less the expansion in the small cylinder, the greater the economy must be.

It has often occurred to ourselves why these simple facts have been so much "passed by," for this reason:—

The high-pressure cylinder is only an *introducer* of the initial steam, and that steam—due to its pressure—requires *time* to allow its elastic force to operate.

Now, if that steam rushed *at once* into a cold atmosphere—as it is termed—a sudden losing change must occur, but if time is allowed and a gradual reduction of temperature permitted also, then certainly the elastic force will best expend itself.

Obviously, then, to expand steam economically time must be considered as analogous with pressure and temperature.

We now come to the operation of the steam and its effect in the low-pressure cylinder after the duty performed in the high-pressure cylinder.

To a certain extent it might be said we have all the evils to overcome again, but with this difference: the low-pressure cylinder having an increased area, the piston has only to move a short distance from the end of the stroke to enable all the *reserved* bulk of steam to fill the vacancy.

This bulk of steam, it must be remembered, is direct from the receiver and valve casing, combined, also, with the steam on the exhaust side of the piston of the high-pressure cylinder. Now, this steam is not of such very much less temperature than what will follow it—as before explained—and the result is that the low-pressure cylinder is *warmed* by the "reserve" steam sufficiently to receive the next "new" steam without such loss as would occur in a single cylinder.

We may here observe that the advocates of single cylinders for the full expansion of steam seem to have forgotten—*i. e.*, if they knew—that cooling means liquefaction, and liquefaction results in reduction of elastic force, causing, therefore, a waste of available power.

But there is another fact that we must point out—which is also generally overlooked: we mean the back pressure in compound-engines, or, rather, the effect of the "reserved" steam between the two cylinder pistons.

We must next suppose the low-pressure piston to be at the end of its stroke while the high-pressure piston is at about half-stroke.

The steam, then, that is exhausting at the back of the high-pressure piston has a certain pressure, and that pressure acts against the piston while it moves on, and, in fact, is acting against the piston until the cylinder is empty.

The practical result is, that the steam will not leave the cylinder unless there is sufficient space to enter into beyond, and that is the reason why receivers were introduced with compound-engines.

It was to prevent cooling in the "receiver" that Cowper introduced his "hot-pot," in which he reheated the "reserved" steam, and also the succeeding volume. The reason why Cowper "reheated" the steam was to restore its elastic force.

The final effort of the steam in the low-pressure cylinder has next to be explained. On the valve opening the exhaust port, the exhaust steam "rushes" to the condenser, because a vacuum is formed, and thus a "space" admissible for the steam to enter.

Obviously, therefore, there is no back pressure in the low-pressure cylinder with a condenser. We mention this fact as an axiom for the benefit of young engineers.

CHAPTER III.

RELATIVE POSITIONS OF THE CYLINDERS OF COMPOUND-ENGINES.

VERTICAL MARINE ENGINES.

ALLIBON.—For each crank pin. High pressure cylinder over the top of and partly recessed in the low pressure cylinder. Separate slide valves at the sides. Receiver pipe vertical at the back.—Date 1870.

BURGH.—For each crank pin. High pressure cylinder within the length of the low pressure cylinder. Annular trunk piston. Single slide valve at the sides. No receiver.—Date, 1859.

ELDER.—For three crank pins. High pressure cylinder at the side of small low pressure cylinder, and it at the side of a larger low pressure cylinder. Three slide valves: first, outside the high pressure cylinder; second, between the high and small low pressure cylinder; and, third, between the two low pressure cylinders. No receiver.—Date, 1862.

General Use.—For two crank pins. High pressure cylinder at the side of the low pressure cylinder. High pressure slide valve outside the cylinder. Low pressure slide valve between the two cylinders. Receiver surrounding the high pressure cylinder.

Another Arrangement.—High pressure cylinders as before. Separate slide valves fore and aft at the sides. Receiver surrounding both cylinders, or receiver pipes on each side port and starboard.—Dates, 1865 to 1876.

Howden.—For two crank pins. High pressure cylinder at the side of the low pressure cylinder. Single slide valve between the two cylinders. Receiver between the valve and the high pressure cylinder.

Another arrangement for one crank pin. High pressure cylinder below the bottom end of the low pressure cylinder. Single slide valve at the side of the low pressure cylinder. No receiver.—Date, 1862.

Inglis.—For two crank pins. High pressure cylinder at the side of the low pressure cylinder. Three Corliss valves top and bottom, between and outside both cylinders. No receiver.—Date, 1868.

Macnab.—For two crank pins. High

pressure cylinder at the side of the low pressure cylinder. The slide valves between the two cylinders. Receiver between the valves.

Another arrangement for three crank pins. Two high pressure cylinders, port and starboard; two low pressure cylinders, fore and aft. Three slide valves between the high and low pressure cylinders. Receiver surrounding the high pressure cylinder.—Date, 1860.

PERKINS.—For one crank pin. High pressure cylinder on the top of low pressure cylinder. Piston fitted with trunk to fill up high pressure cylinder. High pressure cylinder, single acting. Low pressure cylinder, single acting. Valve fore-and-aft of cylinders. No air pump. Surface condenser. No receiver. Steam pressure, 300lbs. on the square inch.—Date, 1870.

ROWAN.—For each crank pin. High pressure cylinder on the top of low pressure cylinder, the latter having a piston trunk in it. Slide valves arranged as Howden's or separate valves and a receiver.—Date, 1860.

STEWART.—For three crank pins. High pressure cylinder with steam ports central of the length at the sides. Low pressure cylinders, one on each side of the high pressure cylinder.

Separate low pressure slide valves between the high and low pressure cylinders. High pressure slide valve at the back of the high pressure cylinder. No receiver.—Date, 1873.

HORIZONTAL MARINE ENGINES.

ALLAN.—For each crank pin. High pressure cylinder formed by a cylindrical box having a piston at each end, contained in the low pressure cylinder, thus forming an annular space. The low pressure cylinder has an annular piston enclosing the box. Two slide valves, connected, on the top side of the low pressure cylinder. No receiver —Date, 1862.

COWPER.—For two crank pins. High pressure cylinder at the side of low pressure cylinder. Separate slide valves fore and aft, at the outsides of each cylinder. Receiver a "hot pot" or a jacketed cylinder under the two engine cylinders.—Date, 1863.

DUDGEON.—For each crank pin. High pressure cylinder within low pressure cylinder. Slide valve at the side of low pressure cylinder. No receiver.—Date, 1864.

GENERAL USE.—For two crank pins. High pressure cylinder at the side of low pressure cylinder. Separate slide valves at the side of each cylinder fore and aft. Receiver surrounding both cylinders.

Another arrangement for each crank pin. High pressure cylinder formed by a piston having a trunk, thus forming an annular space for the initial steam. The low pressure cylinder containing the piston and trunk. Slide valve at the outer side of the low pressure extended. No receiver.—Date, 1865 to 1876.

HUMPHREY.—For each crank pin. High pressure cylinder at the back of low pressure cylinder; the latter contains a piston having a trunk at the front side, for the connecting rod to work in. Separate slide valves at the outer side of each cylinder. Receiver or pipes as required.

Another arrangement.—Two high pressure cylinders at the front end; above and below the main piston rod of the low pressure cylinder, the piston of which has three rods. Slide valves at the sides of each cylinder. Receiver or pipes as required.—Date, 1859.

MAUDSLAY.—For each crank pin. High pressure cylinder, half recessed in the front end of the low pressure cylinder. The piston and back end cover being recessed to correspond.

High pressure slide valve on the top of cylinder and the low pressure valve at the side of cylinder. No receiver. Date, 1870.

PENN.—For two crank pins. High pressure and low pressure cylinders side by side, both having double trunks. Separate slide valves at the sides of the cylinders fore and aft. Receiver, surrounding the cylinders.—Date, 1870.

SCOTT or RENNIE.—For each crank pin. High-pressure cylinder on the top of low-pressure cylinder. The pistons, connected by a vertical beam or lever. Separate slide valves at the sides of each cylinder. Receiver, either a separate box, or surrounding the high- and part of low-pressure cylinder, as required.—Date, 1868.

OSCILLATING MARINE ENGINES.

GENERAL USE.—For two crank pins. High and low-pressure cylinders, separate in motion, side by side. Separate slide valves at the sides of each cylinder. Receiver, inner trunnions, and a box, if required.—Date from 1869 to 1876.

GLANVILLE.—For one crank pin. High-pressure cylinder within the low-pressure cylinder, thus forming an annular space for the expansion of the steam. Separate horizontal slide valves at top and bottom of high-pressure cylinder. Vertical slide valves at side of low-pressure cylinder. Receiver pipes leading from valve casing to top and bottom of high-pressure cylinder.—Date, 1863.

BEAM LAND-ENGINES.

EARLE.—For one beam-pin. High-pressure cylinder at bottom of low-pressure cylinders. Plug-valves at each end of both cylinders. Receiver pipes.—Date, 1805.

HORNBLOWER.—For one beam pin. High-pressure cylinder at the side of low-pressure cylinder, inside, or next to beam. Cylindrical piston valves. Receiver surrounding-valves.—Date, 1781.

HAERLEM CORNISH ENGINE.—For one beam pin. Low-pressure cylinder surrounded by low-pressure cylinder. The high-pressure is open at the top or common to the low cylinder. The
of the low-pressure cylinder is open to the condenser. Double-seat valves arranged top and bottom, outside of low-pressure cylinder.—Date, 1846.

MacNaught.—For two beam pins. High-pressure cylinder within crank-pin circle. Low-pressure cylinder at the opposite end of beam. Receiver pipe.—Date, 1847.

Simpson.—For one beam pin. Very similar as Hornblower's arrangement, the improvement being in the reduction of the receiver, and piston valve closer to the cylinder ports.—Date, 1868.

Sims.—For one beam pin. High-pressure cylinder on the top of low-pressure cylinder. The steam acting on the top and bottom only of the respective pistons, and a vacuum forming between the pistons; being, in fact, a single cylinder of unequal diameters.—Date, 1840.

Whittle.—For one beam pin. High-pressure cylinder (open at the top or bottom, as required) enclosed in low-pressure cylinder. Separate slide valves at the sides of low-pressure cylinder. No receiver.—Date, 1868.

HORIZONTAL LAND ENGINES.

Adamson.—For three crank pins. High-pressure cylinder at the side of two low-pressure cylinders. No receiver.—Date, 1875.

Delany.—For one crank-pin. High-pressure cylinder at the back of low-pressure cylinder.

Separate slide valve at the top or on the sides of each cylinder. Receiver pipe.—Date, 1863.

FAREY.—For one crank pin. High-pressure cylinder at the front of low-pressure cylinder. Separate slide valves at the top or on the sides of each cylinder. No receiver.—Date, 1867.

GENERAL USE.—High-pressure cylinder at the side of low-pressure cylinder for two crank pins; while for one crank pin the high-pressure has been put on the low-pressure cylinder, or at its back or front.—Date 1856 to 1876.

CHAPTER IV.

HOW TO DESIGN A COMPOUND-ENGINE.

CYLINDERS AND VALVES.—Having settled the diameter of the cylinders for high and low-pressure steam, next settle the length of crank-pins, width of cranks, and length of intermediate shaft-bearing, because those dimensions determine the distance between the centres of the cylinders, to a great extent, particularly in moderate-sized engines.

The high-pressure cylinder, in most cases, has what is termed a "liner." This liner is sometimes made of cast steel, while in other cases cast iron has been used; but we recommend wrought iron, having used it for running at high speeds, and steam pressure at 200lbs. on the square inch.

The high and low-pressure cylinders should, in most cases, be in separate castings for the reason of manufacture and repair.

It is the fashion in these days to admit the liner in the high-pressure cylinder only, but we recommend a liner to be used in the low-pressure cylinder also.

The advantage of liners is twofold: first, the liner can be made of any material best for its purpose; second, in the case of repair—or, rather, re-boring—the liners can be removed from the cylinders, and the latter remain in the ship. The securing flange for the liner should always be at the bottom or front end of the cylinder, for the practical reason that the piston is removed or put in from the top or back end.

The space between the liner and the casting, or enclosing cylinder, may be used as a steam-jacket.

The position of the slide valves depends very much on their design; but it must always be remembered that the nearer the valves are together the less distance the expanding steam has to travel—the receiver being, of course, in capacity to the high-pressure cylinder.

We find in most modern engines, particularly the vertical kind, the low-pressure valve is between two cylinders, and the high-pressure valve at the side of its cylinder.

For horizontal engines it is more the fashion for the valves to be fore and aft of the cylinders.

We recommend in all cases cylindrical valves with an expansion cylindrical valve within the larger or exhaust valve. We have worked those valves for some time with great success. They are entirely equilibrium: as, for example, independently of the weight of the valve, the is so little that a valve nine inches in diameter when under steam can be moved direct with one hand.

With large engines the high-pressure cylinder should have two valves at its outer side.

The low-pressure cylinder should have four valves of the same diameter as the high-pressure cylinder. Two valves between the two cylinders and two valves at the outer side.

The reason for this is that the same pattern

for one valve is equally available for the remaining five. In some cases, however, it is equally wise to have two valves only sufficiently large for the low-pressure cylinder, and thus save one set of link and expansion motions at the cost of new patterns for the larger valves.

When flat side valves are used they can be double or single ported, with "plate" valves at the back; while in some examples a "gridiron" valve is used at the back of the low-pressure valve.

The bottom, or front end of the cylinder, should always be made with a large boring-hole that is fitted with a cover, containing the stuffing-box of the piston-rod.

The top or back end cover is very often similarly fitted, but contains the relief valve, surmounted with the usual spring, set screw, and casing.

The bottom or front end relief valve being at the bottom side, or thereabouts, of the inside diameters of the cylinders.

We next come to the question of steam jacketing, not as to its advantages now, but rather as to its construction.

We advise steam jacketing by all means; and in its construction put in as many connecting ribs as the moulder can practically make.

Divide the ribs equally; leave spaces sufficiently

to take out the "core." Avoid sharp corners, and remember one great fact, in particular, that steam jacketing proper is that the sides, the ends, and, in fact, the entire surface, of the cylinders and valve casings must be covered, or enclosed with steam; therefore, if any portion is not jacketed, a great deal of heat will radiate from that surface.

The stuffing boxes for the piston and valve rods should be fitted with deep "bushes" grooved on the inside to contain water or oil, to prevent the steam passing.

Besides those deep bushes it is desirable to put in a loose-grooved bush, centrally, of the depth of the packing, to intercept any steam that might pass the bush.

Further than this the gland is sometimes grooved.

Glands in all cases should be shifted by worm-wheel gearing.

Blow through-holes should be at the ends of each cylinder.

The indicator holes should be at the ends of the cylinders as near the centre as practicable.

Going back again to relief valves, it sometimes happens that it is wise to have a safety valve on the receiver, which should be fitted also with blow-out holes. A man-hole is sometimes added.

In the event of either engine breaking down, slide stop valves are fitted to the exhaust passage metal, which can be opened or closed, to suit the requirements of the case.

As the cylinders contain the pistons we may as well explain them here.

The most modern used piston is cast hollow, with radial ribs. The top, or back portion of the piston is generally curved.

The rod passes through a "boss," and is secured by a nut at the back.

The flange and spring rings are of the usual kind. The spring ring is packed at the back with a series of curved springs. The flange ring is secured by a series of studs, that screw into recessed nuts in the body of the piston.

We now conclude this treatise on the cylinders and their appendages by stating that *all* the steam jacketing must be well lagged with felt and wood.

SURFACE CONDENSER.—This portion of the marine engine has now pretty well settled itself into one arrangement—that is, with the tubes

circulating water to pass through.

For vertical engines the air and circulating pumps are vertical, as a rule, with the exception when the "motion" is taken from the crank shaft or crank pin, then the pumps are horizontal.

The vertical pump is worked by a lever, connected by a link; the other end of the lever is connected by a link, also, to the guide-block of the engine piston-rod.

The advantage of the vertical over the horizontal position is, that the condenser can be lower in the hull. But we are inclined to think that is not of much importance, because there is plenty of room between the under side of the cylinder and the floor line for the condenser to be raised.

With reference to the two "motions," the horizontal has the least amount of detail to look after, but it "breaks up" the proper order of the bearings of the cranked shaft, particularly when the excentrics are secured to the sides of the cranks, or between the inside bearings.

In arranging the valves for the air pumps for horizontal action, be careful that *all* the valves, both for suction and discharge are above the pump.

Vertical pumps are usually single action, the suction valves being at the bottom of the barrel, the delivery valves on the piston of the pump, and the discharge valves at the top of the barrel: but those valves can be arranged at the side of the pump, if desirable—the advantage gained being that the intermediate or piston valves can be dispensed with, while, also, the suction and discharge valves can be "got at" by separate doors.

We have in two or three cases arranged the suction and discharge valves above the pump, causing thereby a considerably better vacuum with the suction valves inverted; in fact, those valves should always be inverted.

For horizontal engines the motion for the air and circulating pumps are generally taken from the steam piston or its rod, with direct-acting engines.

When the return connecting-rod type is used the motion for the pump is taken from an "arm," keyed on the lower piston rod.

We may here remark that the return-action type is dying an easy, natural death comparatively.

Alluding, next, to the condenser tubes, the best method for packing them is the tape or other packing with the screwed gland.

When large condensers are required the tubes must be grouped, so that the doors are not too large for taking "off and on."

Bottom "blow out" and "snifting" valves should be as low in the condenser as possible.

With reference to the arrangement of all types of engines herein mentioned, complete access to all the doors must be the main consideration.

FEED AND BILGE PUMPS.—The motion for those pumps in nearly all cases is obtained from the same source as for the other pumps.

The arrangement of the valves for horizontal action is at the gland end of the barrel, so that the air shall escape at each stroke.

The same result is obtained with vertical pumps by putting the valves at the top of the barrel.

The stuffing-boxes and glands should be as for the steam-rods.

LOWER-MAIN FRAME.—This frame is of cast iron for small engines, and is in one casting; but for large engines in two or more castings bolted together.

We prefer to cast it in duplicate halves, connected across the hull between the inner bearings, the caps for which should be of wrought iron, and the brasses "square" in their seats, the frame should be well ribbed and flanged, and the holding-down or securing-bolts distributed throughout the flanges. The whole of the connecting surfaces should be raised for planing, and "key-ridges" should be at each end or sides for proper adjustment.

Those frames are usually made "box-girder"

section under the bearings, and also beyond them; and common with that the entire frame is made hollow in some cases.

We have known these frames to be of wrought iron constructed by plates and angle iron; and there is not the least doubt, when the least weight is required compatible with strength this material supersedes cast iron.

Cylinder Supports.—These supports are of two kinds for vertical engines: one kind is an angular box pillar, two under each cylinder port and starboard

The inner side is vertical for the length of the stroke of the piston, plus the length of the guide block; this makes a substantial support, but at the same time is very heavy, very cumbersome, and very ugly.

To eradicate those three faults, the other kind of support is sometimes cast with the condenser, as also is the guide for the guide block.

The support on the opposite side is a plain wrought iron pillar, secured in most instances by nuts at each end. This is the most modern arrangement, and certainly looks better, although, perhaps, not so "solid" as the box kind.

Sometimes the flanged girder section is used.

Main Frames.—Those frames are used for

horizontal engines, are generally of the > shape, of cast iron, with ribs, and holding-down flanges.

The cylinder end of the frame is secured to suitable projections cast on the cylinder. The crank shaft end is secured to the condenser near the base line.

The brasses, caps, and securing bolts are much the same as for vertical engines.

We have known horizontal frames to be cast hollow with a plain exterior, and the upper part secured to the cylinder by a wrought iron stay-rod—that is, a prolongation of the upper securing bolt of the cap.

We have known, also, those frames to be of wrought iron, constructed by plates and angle iron. They are much lighter but more susceptible to vibration, or tremor, than cast iron, while the cost of manufacture may be said to be more also.

GUIDE BLOCKS.—These are of two kinds, double and single flanges.

The adjustment of those flanges is either by wedges, with screws and nuts at each end, or by an inclined surface and set studs at one end.

The double-flanged guide block is used for port and starboard surfaces, most often when the box "supports" are used.

The single-flanged block is often termed the "slipper" guide block.

The guide is either cast with or bolted on the condenser, and has a broad channel centrally throughout its length. The guide is formed on each side by guide flanges forming two upper surfaces, the under or back surface being plain

The wearing surfaces of the guide block are the whole area at the back or bottom, while the upper surfaces extend only on each side the connecting rod.

We have introduced, with great success, combined with lightness and economy, four guide rods of equal diameter. The block is of a cross form in plan, and the rod-caps were adjusted by studs.

The most modern method of connecting the piston rod with or to the connecting rod is for the latter to clasp the block. The connecting pin passes between two "eyes" and the block.

The piston rod has a ⊥ end, the flange part of which, is secured to the block by bolts that pass through it and a cap over the connecting pin—nuts and stop studs making good the connection.

Main Connecting Rod.—The portion of this detail that we shall now refer to is the crank pin end, which is of two kinds.

The first is a solid end in one forging, drilled out centrally to receive the brasses, divided across and fitted with securing bolts and nuts, the bolts being so near the crank pin that the brasses are grooved, and, being circular, are thereby prevented from shifting.

The second kind is semi-square, with flat ends for the ⊣ end of the rod, and the cap to fit against.

The securing bolts pass through the four details, and are fitted with nuts and heads as usual.

We shall, of course, be expected to explain which is the better of the two kinds. The flat brass kind is subject to one great fault, and that is the ⊣ end of the rod is the only portion, and, being shallow, that supports the securing bolts. But with the solid head kind one-half of each bolt is contained in the head, and the cap contains the other half.

There is also another feature, with a difference, in the two kinds, relating to the brasses.

With the flat brass kind the brasses are held in lateral position by the securing bolts only.

But with the solid head kind the brasses are flanged, and, therefore, a more perfect "seating" is obtained.

SECURING BOLTS AND NUTS.—The portion now

to be explained, refers to the best means to prevent the bolt and nut from becoming loose when they are in motion.

When it is remembered that those details resist the whole of the power of the engine, and are also in motion and subjected to a series of changing strains—that is, torsion, compression and tensile—it can easily be understood they require some attention.

The best means to prevent the bolt from becoming loose is, set studs at each end, just within the head and behind the nut.

The best means to prevent the nut from becoming loose is to make a certain portion of it circular. Recess that portion into the cap, and at the side insert a set stud.

With very high velocities it is better to have a split-pin passing through the end of the bolt beyond the nut, while in other cases the nut has a series of cross grooves for the split-pin to fit in.

LINK MOTION.—The main portion of this subject that requires our explanation is the form of link that will give the least up-and-down motion during its traverse forward and backward, or, what is known amongst those who understand it, the least amount of "lost motion."

It must, of course, be understood that the

extremities of the link where connected by the excentric rods, form loops or ovals for the centres of motion.

Now, the narrower those loops are, the lesser the "up-and-down" motion, and at the same time a reduced "lost" motion.

We may here explain that much of this motion is due to the way—or, rather, the point—from which the link is hung, it being understood that the length of the suspension rod describes an arc, because it is moved by the link, while the link is raised and lowered the depth of the versed sine of the chord of motion.

The link that will give the least lost motion is the twin solid bar link, with the excentric rod rod pins, and the suspension pins on the same curve centrally of the width of the link, because it has only one motion at each end.

The next best link and most popular is the slotted link with the excentric rods attached beyond the inside curve.

But this link gives four motions, two at each end—that is, the excentric pin motion and the motion at the block pin. The block pin motion is a very narrow ellipse, while the excentric pin motions are very broad ellipses.

The best point to suspend this link from is at the centre of its slot.

From the preceding remarks it can be readily gathered that the "twin bar link" has the least motion of the two examples; but we are not at all complaining of the slotted link, beyond its two motions, because we know, from practice, that by allowing the upper part of the link to "rest" on the block the excentric pin motion at that part is much reduced in width.

The main feature to be considered with the link motion is its facility for starting, stopping, and reversing engines.

The link, then, that can be raised up and lowered down with the least amount of friction must be the best, and it is but fair to state that for that purpose the two links we have quoted are equal.

Regarding the difference in the lost motions again, we may here observe that in both cases the lower excentric pins but very little affect the motion of the upper pins, and *vice versa*, because the suspension pin intervenes.

EXPANSION GEAR.—The use of this gear is to govern the expansion valve, so that it will intercept or "cut off" the steam at the back of the main, or exhaust, valve at any required grade of expansion.

This gear, also, must be so arranged that it can

be altered to alter the grade of cut-off whilst the engine is in motion. It must be so arranged, also, that the expansion valve can remain stationary.

The best mechanism known in the present day is a slotted curve linked on a shaft at the end; worked by an excentric and rod, the link being fitted with a sliding block that is attached to a rod connected to the valve rod.

The shifting of the block is accomplished either by a screwed rod or lever.

Many arrangements have been made to accomplish this, but the most simple and effective is that we have described.

STARTING, OR REVERSING, GEAR.—This gear is used to shift the link forward and backward, or keep it stationary in any point. The best arrangement yet known is the sliding block and screwed rod, on the end of which is the hand wheel.

Levers may be used, or rods, direct to the link. But for the expansion valves we have explained, the screw block with direct rods will be sufficient.

CHAPTER V.

HOW TO INDICATE A COMPOUND-ENGINE.

WE have explained, in Chapter II., the science of indicating the action of the steam; and our purpose now is to explain how to do that practically.

If we suppose ourselves looking at the "line" of motion of steam when doing its duty in the cylinder, we shall see that the steam has a tendency to force off the cylinder ends, so that the "line of motion" may be said to be from end to end.

The indicators, therefore, ought to be fitted to the front and back ends of the cylinders.

It is too often the practice with horizontal cylinders to fit the indicator vertically; but this should. never be, for the reason that if the steam has to travel through a bend-pipe an equivalent amount of friction must result, and the pressure of the steam lowered thereby.

We therefore give this axiom:—Secure an

indicator at each end of each cylinder. If the cylinder be horizontal let the indicator be horizontal; and if the cylinder be vertical let the indicator be vertical.

The next particular matter to be observed here is the arrangement of the indicator gear.

In all cases obtain the motion from the piston rod; avoid more than one lever for each indicator, if possible; but with long-stroke engines, at the back end, an angle lever may be used, in order that the string may be as short as practicable, it being known, from practice, that a long string allows a "loop" motion.

With regard to the length of the indicator diagram, about five inches is the usual limit. But with engines running at high velocities—say, from 350 to 500 revolutions per minute—four to three inches are the better lengths.

The reason for this difference in the length of those diagrams is, that the marker must have sufficient time to record its motion in conjunction with the motion of the barrel.

One of the worst practices in the present day is to indicate the steam in the cylinder of compound engines separately, whereas each end of the cylinder should be indicated at the same moment by an operator situated at each instrument.

One end of the high-pressure cylinder should be indicated, and at the moment of the return stroke the opposite end of the low-pressure cylinder should be indicated: this is theoretical, as far as comparison is concerned; but we find in practice that a sufficiently truthful result is obtained by indicating at each end of the cylinders at the same time.

The mechanical operation of indicating an engine is as follows:—Having the "card" and barrel in their proper place, attach the string to see that all is right as regards the motion and springs.

The indicator should now be well "blown through," the steam shut off, the water allowed to drain out of the indicator, and, next, the atmospheric line taken. The steam from the cylinder should now be allowed to enter the indicator again, the word "ready" should be given, each marker "handed" delicately by each operator, and the diagrams taken from each end of the cylinder by the closing and opening the respective cocks.

The following notes should always be taken and marked on each diagram.

Low or high pressure cylinder.

Back or front, top or bottom end.

Pressure of steam shown by gauge in engine room.

Pressure of steam shown by gauge in boiler room.

Vacuum shown by gauge on condenser.

Number of revolutions per minute.

Date and time of taking of the diagrams.

The notes to be taken in the pocket-book are a very great deal according to circumstances of the required amount of information, but it is usual to note the following:—

Name of the ship.

Estimated nominal horse power of the engines.

Diameter of each cylinder.

Length of the stroke of the piston.

Travel of the main slide valve } of each cylinder.
Travel of the expansion valve }

Length of crank connecting rod.

Amount of coal consumed per hour or per half-hour in pounds.

Amount of ashes as refuse per hour or per half-hour in pounds.

Speed of the ship in knots per hour.

Force of the wind.

CHAPTER VI.

THE ANALYSIS OF THE INDICATOR DIAGRAM.

THE vertical line from the atmospheric line is the "admission line."

The horizontal line parallel with the atmospheric line at the top of the admission line, is the "initial supply line."

The curved line leading from the last point is the "expansion line."

The line from that point that curves downwards towards the bottom line of the diagram is the "initial exhaust line."

The bottom line that is parallel with the atmospheric line is the "final exhaust line."

The curved portion that joins the end of the preceding line with the admission line is the "compression and lead line."

There are, therefore, six lines formed by the marker of the indicator that relate to the action of the steam in the cylinders during one revolution of the crank pin.

We have arranged their description, therefore, in the following order:

1st.—Admission line.
2nd.—Initial supply line.
3rd.—Expansion line.
4th.—Initial exhaust line.
5th.—Final exhaust line.
6th.—Compression and lead line.

It will be noticed that we have supposed the two diagrams, that is from the high and low pressure cylinders, to be of the same scale "pieced" together, which of course they should be, to show their relation to each other.

We may here explain that the compound-engine, although it may have two or three cylinders, as the case may be, it is practically in theory as a single cylinder engine; and the only reason why the extra cylinder or cylinders besides the high pressure cylinder is used, is, as we said before, to allow "time" for the elastic force in the steam to expend itself.

Therefore, when the diagrams do not piece together well, that is, when the exhaust line of the high pressure diagram does not form the same shape as the admission line of the low pressure diagram, or in other words, one line will not "record" on the other, we know there is some mistake.

Now, the next question, therefore, is, what is the mistake and the cause thereof?

The main defect will proceed from the bad arrangement of the moving steam valves and their proportions, while at the same time a little fault often occurs in the diagram not being taken properly. It has often occurred to ourselves that the portion of the low pressure diagram that is above the atmospheric line may be considered as "back pressure," while at the same time the distance, should there be any, between the two lines of contact of the two diagrams must be considered as back pressure also.

It is the practice in some cases to "scale" the diagram separately above and below the atmospheric line, while in other cases the entire area of the diagram is taken as one.

The best method to obtain the area of a diagram is to divide its length by ten equidistants setting off one-half space at each end, the sum of each line forming a column outside the diagram. The whole of the sums added together divided by the number gives the mean pressure.

We have considered the indicator diagram thus far as being a practical illustration of the action of the steam in the engine cylinder, but what we have to further consider chiefly is, the difference between the diagram in theory and the diagram in practice.

The theoretical diagram shows a vertical rectangular figure, or it may be a square, for the area of the initial steam portion.

The expansion portion is a right angle figure, joined by a hyperbolical line; the hyperbolical curve being admitted to be the theoretical curve described by the marker as it descended, and the barrel turning around during the expansion of the steam.

The exhaust portion is a horizontal rectangular figure with a concave corner.

The remaining portion of the diagram is the compression and lead in practice, but should be omitted according to theory.

We will now explain the difference between the theoretical and practical diagrams.

The initial steam line is very often angular downwards instead of being horizontal; it is connected with the expansion line by a convex curve, while the concavity of the expansion curve is seldom down to the hyperbolical line. The descending portion of the exhaust line is always convex, whereas theory gives it as concave. The

bottom corner is very often knocked off.

We, of course, shall be expected to explain the

result of this difference, which is that the area of the diagram in practice is much larger than in theory; and when we think it over calmly, it is easily to be accounted for.

The initial steam line, although angular in practice, does not compensate in loss for the fully developed expansion line, which development is further carried out in the descending exhaust line.

To make matters balance, therefore, it is wise to consider all the back pressure shown in the diagram, and to deduct the area of that back pressure from the areas of the diagrams; because if not so done the mean pressure shown as recorded will include the back pressure and therefore the indicated horse power will be made out to be more than it really is.

To better impress those facts we will again explain what back pressure is, as shown by the indicator diagrams. It is that portion above the atmospheric line that is between it and the exhaust line of the high pressure cylinder. It is also the whole of that portion between the two diagrams when they do not piece together.

We are under the impression that the back pressure is too often "smuggled" in with the actual pressure, and thus the latter is said to be more than it really is.

CHAPTER VII.

THE VALUE OF A UNIT OF HEAT IN STEAM IN COMPOUND-ENGINE CYLINDERS.

PRINCIPLES.—We have now to discuss the formation of a unit of heat; the sum of that formation, and, lastly, the value of it.

To begin with, we must consider the names of the proportions, which are these:

1.—Pressure in lbs. per square inch.

2.—Sensible temperature of degrees Fahrenheit.

3.—Total heat in degrees from zero of Fahrenheit.

4.—Weight of one cubic foot of steam.

5.— Relative volume of the steam compared with the water from which it is raised.

The formation of a unit of heat in steam is the amount of heat absorbed by the water and air proportionately, and the form or shape of that unit is a series of globules which contain the heat that set them in motion.

It therefore only requires a knowledge of the

constituents and their proportions to determine what sum a unit of heat in steam is.

Now let us go through this matter carefully. We know that the quantity or cubical contents of the initial steam in the high pressure cylinder is the motive power for one revolution of the crank pin in a two or three cylinder engine. Our next consideration, therefore, is what that volume of steam contains, and to what limit can we use it.

The contents of that volume we have explained, and our next step is its total value in weight; because the "weight" of the steam that "drives" the engine is really the bulk of power expressed by that term.

The weight being in proportion to the pressure, and the pressure being in proportion to the temperature, pr es that after all we must come back again to theoword "heat." Consequently we have to consider two great facts in finding out what a unit of heat is.

First, the weight of the whole of the constituents in the initial steam.

Second, the sensible temperature in "foot" degrees that produced that weight. By which means the number of units of heat in that initial steam is obtainable.

These conclusions, as far as we know, are original, for the reason that we have not dealt with one supposition but with facts only.

In stating that, we are not ignoring Joule's resultant, which has been acknowledged as 1390 units of work to represent a unit of heat, which is called 772 foot pounds, for each unit of heat. So that if the units of work are divided by the foot pounds, we are said to have produced the number of units of heat required. It may be well to add that this unit of heat is in a normal condition, because the unit is considered as the value of one pound of water with one degree of heat Fahrenheit in it.

We learn, therefore, that so many pounds of water and so many degrees of temperature in equal numbers represent so many units of heat, and those units multiplied by the constant 1390, represent the work performed. And we suppose that if the work performed in one minute is divided by 33,000 —Watt's constant for one actual horse power—the power of those units of heat will be known.

It is worth while explaining that the method Joule employed to arrive at his constant 1390 was that the effort required to raise one pound of water at one degree of temperature one foot high was equal in effect to the falling of one pound weight 1390 feet deep. We, however, offer no opinion on that point, but rather prefer to consider the weight and temperature of the steam, as shown by the indicator diagram.

Our reason for this is, that if we use the indicator as an instrument to show what steam has been used as initial steam, we must make use of the diagram shown as our standard, to arrive at the amount of heat used per revolution of the crank pin.

Now then, suppose we have the horse power given to us, we have only to work the matter backwards in calculation and we shall obtain the cubical contents of the initial steam, from which we get the length of the "cut off" and area of the high pressure cylinder.

The formulæ for this will be found farther on.

CHAPTER VIII.

THE LOSS OF THE HEAT IN THE STEAM IN COMPOUND-ENGINE CYLINDERS.

Principles.—In dealing with this matter we shall, as in the preceding chapter, explain the basis of its standing.

Now let us suppose an engine has given out a certain indicated horse power, as shown by the indicator diagrams.

The contractors are contented, the owner is satisfied, and the engineers of the ship consider themselves fortunate in their appointments.

But suppose we disturb all this equanimity by stating that according to the indicator diagram the engine is a mistake, because it consumes double as much steam as it ought to in practice and two-thirds as much in theory—we shall be asked, of course, to prove this—the principles of which we now explain.

Starting then with the indicated horse power, we have also the following particulars:—

Speed of piston in feet per minute.

Mean pressure in high pressure cylinder }
Mean pressure in low pressure cylinder } in lbs. per square in.

Area of high pressure cylinder }
Area of low pressure cylinder } in square inches.

We have now before us the three principal facts that made the indicated horse power what it was.

Then if the indicated horse power be taken as a whole from those facts, we are bound by principle to take them as a combination.

Obviously then the collective mean pressures in the two cylinders are used; and equally obvious the collective areas of the two cylinders are used on equal terms, because they do combined duty.

Thus far we are all agreed in principle, but we find in practice that it is essential to take each cylinder and mean pressure as a combination exclusively; that is to say, the mean pressure of the high pressure cylinder belongs to its area; and the mean pressure of the low pressure cylinder belongs to its area.

But mark this! the two separate exclusions we have just alluded to results in a given speed of piston.

Now if that speed is a resultant, said to be from combined forces, we have a right to consider those forces as combined, and not separately.

The present acknowledged formulæ to obtain the indicated horse power of an engine is thus—multiply the area of the cylinder by the mean pressure of the steam, which equals the pressure-power.

The stroke of the piston in feet multiplied by 2 equals one revolution, and that sum multiplied by the number of revolutions per minute is said to equal the piston speed in feet.

Next, the result of the latter calculation multiplied by the sum of the Pressure-power is said to equal the Foot-pounds power. Then with Watt's constant 33000 as a divisor into the foot-pounds power, we are said to obtain the indicated horse power; from which result financial operations are often agreed on and carried out.

Now, if we consider this matter carefully, the formula that makes a result should take into consideration the principles of the case.

The principles of the case in this matter are, as we said before, the combination of the two facts that produce one result. We allude again to the combined areas of the cylinders, and the combined mean pressures that work in them.

Next, suppose we have the total indicated horse power given to us, and we multiply that by Watt's constant 33000, we shall have the lineal foot-power in pounds.

Then, if we multiply the collective area of the two cylinders by the piston speed in feet per minute, we shall have the lineal foot-area in square inches —both results being on equal terms of value.

Now, if we divide the foot-power by the foot-area, it will give us the mean pressure of the steam that impels the engine at the given speed.

This mean pressure is in fact the collective power required to work the engine as it should be, but we find in practice that the actual pressure required, is often more than three times the theoretical pressure.

We have taken a great deal of interest in this matter, and given it a great deal of consideration during the last five years, and our conclusions to the present are that *two thirds* of the heat in the steam is lost by imperfections of proportions, radiation, and liquefaction. We find, also, that ranging over 50 examples of our most modern compound engines the steam constant of loss ranges from 2·26 to 3·5 as the divisor, for the actual working mean pressure collectively to be divided by, to obtain the theoretical mean pressure that should have driven the engine at the same speed of piston.

Our firm belief is, that this great loss that we allude to, is as much due to the sub-elasticity of the steam as it is to any of the other causes we have mentioned.

CHAPTER IX.

FORMULÆ TO OBTAIN THE VALUE OF A UNIT OF HEAT IN STEAM IN COMPOUND-ENGINE CYLINDERS.

To Find the Proportion of a Unit of Heat to the Total Indicated Horse Power of a Compound-Engine.

Area of high pressure cylinder in square inches, A.

Length of cut-off in lineal inches, O.

Cubical contents of supply steam in feet, F.

Weight of one cubic foot of steam of the initial pressure, S.

Sensible temperature of that pressure in foot degrees, T.

Units of heat, U.

Total indicated horse power, P.

Constant—value, C.

If we wish to put this into proper formula, it must be done thus:

Multiply the area of the high pressure cylinder A by the length of cut-off O. Divide that sum by 1728 I. Multiply the cubical contents F by the weight of one cubic foot of initial steam at that pressure S which equals the weight of the initial steam used for one revolution of the crank-pin B. Multiply that sum by the sensible temperature of the initial steam T, which will equal the number of units of heat U. Then the total indicated horse power P, divided by the units of heat U, equals the initial heat constant—value C.

To put this into condensed formulæ we must arrange it as shown—

$$\left[\frac{A \times O}{I} = F\right] \quad \left[S \times F \times T = U\right] \quad \left[\frac{P}{U} = C\right]$$

We will now direct attention to a reverse matter, that is, supposing we have settled the following:—

Constant value, C.
Indicated horse power, P.
Weight of initial steam, B.
Sensible temperature, T.
Length of cut off, O.

REQUIRED THE AREA OF THE HIGH-PRESSURE CYLINDER? We must arrange the calculation thus:—Divide the indicated horse power P by the constant value C, which will equal the units of heat required U. Divide the units of heat U by the sensible temperature T, which will equal the weight of the initial steam B. Then the weight of the initial steam B divided by the weight of a cubic foot of steam S, equals the cubical contents of the supply steam in feet F. Next, those contents multiplied by 1728 I equals the cubical contents of the supply in inches. Then those contents divided by the length of cut off O, which will equal the area of the high pressure cylinder A.

To put this into condensed formula we must arrange it as shown—

$$\left[\frac{P}{C}=U\right]\ \left[\frac{U}{T}=B\right]\ \left[\frac{B}{S}=F\right]\ \left[\frac{F\times I}{O}=A\right]$$

To make this fully understood we have given the following examples from actual practice, we must explain, that only two decimals are used in U, so that A direct, and A reverse, are a little unlike.

HEAT CALCULATIONS TO FIND THE VALUE OF A UNIT OF HEAT

OF MODERN EXAMPLES OF COMPOUND ENGINES, BY VARIOUS FIRMS.

S. S. "MONGOLIA," MESSRS. DAY, SUMMERS, & CO., SOUTHAMPTON.

48″ dia. = 1809·56 = area of high pressure [cylinder = A
27″ = cut off = O

1266692
361912

I = 1728) 48858·12 (28·27 = cubical contents of supply in feet = F
3456

14298
13824

4741
3456

12852
12096

756

P = Total Indicated H. P.

U = 1115·54) 1258·00 (1·12 = C
1115·54

142460
111554

309060
223108

85952

lb.
·1364 = weight of 1 cubic foot of steam 42·3 lbs. above atmosphere = S
28·27 = F

95 8
2728
10912
2728

3·856028 = B
289·3° = sensible temperature in foot degrees = T

11568084
34704252
30848224
7712056

1115·5489004 = units of heat = U

$\left[\frac{A \times O}{I} = F\right]$ $\left[S \times F \times T = U\right]$ $\left[\frac{P}{U} = C\right]$

REVERSE CALCULATION TO FIND THE AREA OF THE HIGH PRESSURE CYLINDER.

S. S. "MONGOLIA."

```
              P
C = 1·12) 1258 (1123·21 = U
          112
          ---
           138
           112
           ---
            260
            224
            ---
             360
             336
             ---
              240
              224
              ---
               160
               112
               ---
                48
```

```
                  B
        lb.  ) 3·882509 (28·46 = F
S = ·1364    ) 2728
               -------
               11545
               10912
               -----
                 6330
                 5456
                 ----
                  8749
                  8184
                  ----
                   565
```

```
                 U
T = 289·3°) 1123·21 (3·882509 = B
            8679
            -----
            25531
            23144
            -----
             23870
             23144
             -----
               7260
               5786
               -----
               14740
               14465
               -----
                27500
                26037
                -----
                 1463
```

```
          28·46 = F
           1728 = I
         ------
          22768
         5692
        19922
        2846
       --------
O = 27") 49178·88 (1821·44 = A
         27
         ---
         221
         216
         ---
           57
           54
           --
            88
            27
            --
            118
            108
            ---
             108
             108
```

$\left[\frac{P}{C} = U\right]$ $\left[\frac{U}{T} = B\right]$ $\left[\frac{B}{S} = F\right]$ $\left[\frac{F \times I}{O} = A\right]$

S. S. "LADY JOSYAN," MESSRS. DAY, SUMMERS & Co.

```
26" dia. = 530·93  = area of high pressure
             18" = cut off = O      [cylinder = A
          ------
           424744
           53093
          ------
I = 1728) 9556 74 (5·53 = cubical contents of
          8640              supply in feet = F
          ------
           9167
           8640
          ------
            5274
            5184
          ------
              90
          ======

  P = Total indicate H. P.
U = 218·21) 294·00 (1·34 = C
            21821
            ------
             75790
             65 63
            ------
            103270
             87284
            ------
             15986
            ======
```

```
          lb.
       ·1364 = weight of 1 cubic foot of steam
        5·53 = F              at 42·3 lbs. above
      ------                  atmosphere = S
        4092
       6820
      6820
    --------
    ·754292 = B
      289·3° = sensible temperature in foot
    --------                   degrees = T
      2262876
     6788628
    6034336
   1508584
   ----------
   218·2166756 = units of heat = U
```

$$\left[\frac{A \times O}{I} = F\right] \quad \left[S \times F \times T = U\right] \quad \left[\frac{P}{U} = C\right]$$

Reverse Calculation.

S. S. "Lady Josyan."

```
                 P
C = 1·34) 294 (219·4    U
          268
          ---
           260
           134
           ----
           1260
           1206
           ----
             540
             536
             ---
               4
```

```
               U
T = 289·3°) 219·40 (·758382 = B
            202 51
            ------
             16890
             14465
             -----
              24250
              23144
              -----
              11060
               8679
              -----
               23810
               23144
               -----
                6660
                5786
                ----
                 874
```

```
        5·55 = F
        1728 = I
        --------
        4440
       1110
      3885
      555
      ----------
O = 18") 9590·40 (532·8 = A
         90
         --
          59
          54
          --
           50
           36
           ---
           144
           144
```

```
                 B
         lb.
S = ·1364) ·758382 (5·55 = F
            6820
            ----
             7638
             6820
             ----
              8182
              6820
              ----
              1362
```

$\left[\frac{P}{C} = U\right]$ $\left[\frac{U}{T} = B\right]$ $\left[\frac{B}{S} = F\right]$ $\left[\frac{F \times I}{O} = A\right]$

DIRECT CALCULATION.

S. S. "DANUBE," MESSRS. DAY, SUMMERS & Co.

36" dia. = 1017·8 = area of high pressure [cylinder = A
26" cut off = O

61068
20356

I = 1728) 26462·8 (15·3 = Cubical contents of supply in feet = F
1728

9182
8640

5428
5184

244

P = Total indicated H. P.

U = 701·23) 862·30 (1·22 = C
701·23

161070
140246

208240
140246

67994

lb.
·1538 weight of 1 cubic foot of steam at 50·3 lbs. above atmosphere = S
15·3 = F

4614
7690
1538

2·35314 = B
298° = sensible temperature in foot degrees = T

1882512
2117826
470628

701·23572 = units of heat = U

$\left[\frac{A \times O}{I} = F\right]$ $\left[S \times F \times T = U\right]$ $\left[\frac{P}{U} = C\right]$

REVERSE CALCULATION.
S.S. "DANUBE."

$$C = 1\cdot 22) \overset{P}{862\cdot 3} \ (706\cdot 8 = U$$
$$854$$
$$\overline{830}$$
$$732$$
$$\overline{980}$$
$$976$$
$$\overline{4}$$

$$T = 298°) \overset{U}{706\cdot 8} \ (2\cdot 3718 = B$$
$$596$$
$$\overline{1108}$$
$$894$$
$$\overline{2140}$$
$$2086$$
$$\overline{540}$$
$$298$$
$$\overline{2420}$$
$$2384$$
$$\overline{36}$$

$$O = 26'') \ 26611\cdot 2 \ (1023\cdot 5 = A$$
$$26$$
$$\overline{1232} \quad 15\cdot 4 = F$$
$$1078 \quad 1728 = I$$
$$\overline{154}$$
$$\ \ 308$$
$$\overline{\ \ 61}$$
$$52$$
$$\overline{\ 91}$$
$$78$$
$$\overline{132}$$
$$130$$
$$\overline{2}$$

$$S = \cdot 1538) \overset{B}{2\cdot 3718} \ (15\cdot 4 = F$$
$$\text{lb.} \quad 1\ 538$$
$$\overline{8338}$$
$$7690$$
$$\overline{6480}$$
$$6152$$
$$\overline{328}$$

$$\left[\frac{P}{C} = U\right] \left[\frac{U}{T} = B\right] \left[\frac{B}{S} = F\right] \left[\frac{F \times I}{O} = A\right]$$

S. S. "PETER JEBSON," MESSRS. MAUDSLAY, SONS, & FIELD, LONDON.

Dia. = 29″ = 660 inches = area of high pressure [cylinder = A
15″ = cut off = O

```
         660
          15
        ----
        3300
         660
        ----
I = 1728) 9900 (5·7 = cubic contents of supply in feet = F
          8640
         -----
         12600
         12096
         -----
           504
```

```
lb.
·1714 weight of 1 cubic foot of steam at 58·3 lbs. above atmosphere = S
  5·7 = F
 -----
 11998
  8570
 -----
·97698 = B
305·7° = sensible temperature in foot degrees = T
 ------
 683886
 488490
2930940
----------
298·662786 = units of heat = U
```

P = Total indicated H. P.

```
U = 298·66) 619·00 (2·07 = C
            59732
           ------
            216800
            209062
           ------
              7738
```

$$\left[\frac{A \times O}{I} = F\right] \quad \left[S \times F \times T = U\right] \quad \left[\frac{P}{U} = C\right]$$

Reverse Calculation.

S. S. "Peter Jebson."

```
              P
C = 2·07) 619 (299·03 = U
          414
          ---
          2050
          1863
          ----
           1870
           1863
           ----
             700
             621
             ---
              79
             ===
```

```
               U
T = 305·7°) 299·03 (·978181 = B
            27513
            -----
             23900
             21399
             -----
              25010
              24456
              -----
                5540
                3057
                ----
                24830
                24456
                -----
                  3740
                  3057
                  ----
                   683
                  ====
```

```
      5·7 = F
     1728 = I
     ----
      456
     114
    399
    57
    ------
O = 15") 9849·6 (656·6 = A
         90
         --
          84
          75
          --
           99
           90
           --
            96
            90
            --
             6
            ==
```

```
             B
     lb.) ·978181 (5·707 = F
S = ·1714)  8570
            -----
            12118
            11998
            -----
             12010
             11998
             -----
                12
               ===
```

$$\left[\frac{P}{C} = U\right]\left[\frac{U}{T} = B\right]\left[\frac{B}{S} = F\right]\left[\frac{F \times I}{O} = A\right]$$

Direct Calculation.

S. S. "Nankin," Messrs. Maudslay, Sons, & Field, London.

38″ dia. = 1134 = area of high pressure [cylinder = A
24″ = cut off = O

4536
2268

= 1728) 27216 (15·7 = cubical contents of supply in feet = F
1728

9936
8640

12960
12096

864

P = Total indicated H. P.

U = 915·51) 1221·00 (1·33 = C
915·51

305490
274653

308370
274653

33717

lb.
·1869 = weight of 1 cubic foot of steam 65·3 lbs. above atmosphere = S
15·7 = F

13083
9345
1869

2·93433 = B
312° = sensible temperature in foot degrees = T

586866
293433
880299

915·51096 = units of heat = U

$$\left[\frac{A = O}{I} = F\right] \quad \left[S \times F \times T = U\right] \quad \left[\frac{P}{U} = C\right]$$

REVERSE CALCULATIONS.

S. S. "NANKIN."

```
              P
C = 1·33) 1221(918·04 = U
          1197
           240
           133
          1070
          1064
           600
           532
            68
```

```
              U
T = 312°) 918·04 (2·94243 = B
          624
          2940
          2808
           1324
           1248
            760
            624
            1360
            1248
             1120
              936
              184
```

```
 15·7 = F
 1728 = I
 1256
  314
1099
157
O = 24") 271296 (1130·4 = A
         24
          31
          24
           72
           72
            96
            96
```

```
        lb       B
S = ·1869) 2·94243 (15·7 = F
           1869
          10734
           9345
           13893
           13083
             810
```

$$\left[\frac{P}{C} = U\right] \quad \left[\frac{U}{T} = B\right] \quad \left[\frac{B}{S} = F\right] \quad \left[\frac{F \times I}{O} = A\right]$$

DIRECT CALCULATION.

S. S. "TIMOR," MESSRS. MAUDSLAY, SONS & FIELD.

Dia. = 36″ = 1017 = area of high pressure [cylinder = A
22″ = cut off = O

```
      2034
      2034
I=1728) 22374 (12·9 = cubical contents
        1728        of supply in feet = F
        ----
        5094
        3456
        -----
        16380
        15552
        -----
          828

P = total indicated H. P.
U = 719·79) 1234·00 ( 1·71 = C
             71979
            ------
            514210
            503853
            ------
            103570
             71979
            ------
             31591
```

```
        lb.
    ·1804 = weight of 1 cubic foot of above
     12·9 = F        steam at 62·3 lbs.
    -----            atmosphere = S
    16236
    3608
   1804
  -------
  2·32716 = B
    309·3° = sensible temperature in
                foot degrees = T
  -------
   698148
  2094444
 6981480
 ---------
U = 719·790588 = units of heat = U
```

$$\left[\frac{A \times O}{I} = F\right] \quad \left[S \times F \times T = U\right] \quad \left[\frac{P}{U} = C\right]$$

Reverse Calculations.

S. S. "Timor."

```
          P                             U
C = 1·71)1234( 721·63 = U    T = 309·3°)721·63( 2·3331 = B        12·9 = F
         1197                           6186                      1728 = I
         ----                           -----                     ----
          370                           10303                     1032
          342                            9279                      258
          ----                          -----                     903
          280                            10240                   129
          171                             9279                   -------
          ----                           -----          O = 22″)22291·2 (1013·2 = A
          1090                             9610                  22
          1026                             9279                  --
          ----                             ----                  29
           640                             3310                  22
           513                             3098                  --
           ----                            ----                   71
           127                              217                   66
           ====                            ====                   --
                                                                   52
           B                                                       44
     lb.                                                           --
S = ·1804) 2·3331(12·9 = F                                          8
           1804
           ----
           5291
           3608
           -----
           16830
           16236
           -----
             594
           =====
```

$$\left[\frac{P}{C} = U\right] \quad \left[\frac{U}{T} = B\right] \quad \left[\frac{B}{S} = F\right] \quad \left[\frac{F \times I}{O} = A\right]$$

Direct Calculation.

S. S. "Amerique," Messrs. Maudslay, Sons & Field.

```
dia. 41" = 1320" = area of high pressure
     24" = cut off O        [cylinder = A
           ------
            5280
            2640
           ------
I = 1728) 31680 (18·33 = cubical contents
           1728          of supply in feet = F
           ------
           14400
           13824
           ------
            5760
            5184
           ------
            5760
            5184
           ------
             576
           ======
   P = total Indicated H. P.
U = 900·35) 1617·00 (1·79 = C
              90035
            --------
             716650
             630245
            --------
             864050
             810315
            --------
              53735
            ========
```

```
     lb.
   ·1627 = weight of 1 cubic foot of steam
   18·33 = F                at 54·3 lbs. above
  -------                   atmosphere = S
    4881
   4881
  13016
  1627
  -------
   2·982291
      301·9° = sensible temperature in foot
  ----------          degrees = T
  26840619
  2982291
 89468730
 ----------
 900·3536529 = units of heat = U
```

$\left[\frac{A \times O}{I} = F\right]$ $\left[S \times F \times T = U\right]$ $\left[\frac{P}{U} = C\right]$

REVERSE CALCULATION.

S. S. "AMERIQUE."

```
          P
C = 1·79)1617 (903·351 = U
         1611
         ----
           600
           537
           ---
            630
            537
            ---
             930
             895
             ---
              350
              179
              ---
              171
```

```
              B
       lb.
S = ·1627) 2·992219 (18·39 = F
           1627
           -----
           13652
           13016
           -----
             6361
             4881
             -----
             14809
             14643
             -----
               166
```

```
               U
T = 301·9°) 903·351 (2·992219 = B
            6038
            -----
            29955
            27171
            -----
             27841
             27171
             -----
               6700
               6038
               ----
                6620
                6038
                ----
                 5820
                 3019
                 -----
                 28010
                 27171
                 -----
                   839
```

```
       18·39 = F
        1728 = I
       ---------
       14712
       3678
      12873
      1839
      --------
O=24")31777·92(1324·08=A
      24
      --
       77
       72
       --
        57
        48
        --
         97
         96
         ---
         192
         192
```

$$\left[\frac{P}{C} = U\right] \quad \left[\frac{U}{T} = B\right] \quad \left[\frac{B}{S} = F\right] \quad \left[\frac{F \times I}{O} = A\right]$$

Direct Calculation.

S. S. Garonne, Messrs. N. Napier & Sons, Glasgow.

```
60" = dia. 2827·44  = area of high pressure
            24" = cut off O     [cylinder = A
         ________
         1130976
         565488
         ________
I = 1728) 67858·56 (39·27 = cubic contents
          5184           of supply in feet = F
         ________
          16018
          15552
         ________
           4665
           3456
         ________
           12096
           12096
         ========

      P = Total indicated H. P.
U = 1606·59) 2650·00 (1·64 = C
             160659
            ________
             1043410
              963954
            ________
              794560
              642636
            ________
              151924
            ========
```

```
   lb.
   ·1403 = weight of 1 cubic foot of steam
   39·27 = F                 44·3 lbs. above
  ______                     atmosphere = S
    9821
   2806
  12627
  4209
  ________
  5·509581 = B
     291·6° = sensible temperature in foot
  ________              degrees = T
   33057486
   5509581
  49586229
 11019162
 ________
 1606·5938196 = units of heat = U
```

$$\left[\frac{A \times O}{I} = F\right] \quad \left[S \times F \times T = U\right] \quad \left[\frac{P}{U} = C\right]$$

Reverse Calculation.

S. S "Garonne."

```
            P
C = 1·64)2050·00(1615·85 = U
         164
         ----
         1010
          984
         ----
          260
          164
         ----
           960
           820
          ----
           1400
           1312
           ----
             880
             820
             ---
              60
```

```
          U
T = 291·6)1615·85(5·541323 = B
          14580
          -----
           15785
           14580
           -----
            12050
            11664
            -----
              3860
              2916
              ----
               9440
               8748
               ----
                6920
                5832
                ----
                 10880
                  8748
                 -----
                  2132
```

```
       lb.      B
S = ·1408)5·541323(39·49 = F
          4209
          -----
          13323
          12627
          -----
            6962
            5612
            -----
            13503
            12627
            -----
              876
```

```
          39·49 = F
           1728 = I
          -----
          31592
          7898
         27643
         3949
         --------
O = 24″)68238·72(2843·28 = A
        48
        ---
        202
        192
        ---
         103
          96
         ---
          78
          72
          --
           67
           48
           --
           192
           192
           ---
```

$$\left[\frac{P}{C} = U\right] \quad \left[\frac{U}{T} = B\right] \quad \left[\frac{B}{S} = F\right] \quad \left[\frac{F \times I}{O} = A\right]$$

S. S. "JOSE BARO," MESSRS. OSWALD & Co., SUNDERLAND.

35" dia. = 962·11 = area of high pressure [cylinder = A
24" cut off—O.

3848 4
192422

I = 1728) 23090·64 (13·36 = cubical contents of supply in feet = F
1728

5810
5184

6266
5184

10824
10368

456

lb.
·1648 = weight of 1 cubic foot of steam at 55·3 lbs. above atmosphere = S
13·36 = F

9888
4944
4944
1648

2·201728 = B
302·9° = sensible temperature in foot degrees = T

19815552
4403456
66051840

666·9034112 = units of heat = U

P = total indicated H. P.

U = 666·9) 752·0 (1·12 = C
6669

8510
6669

18410
13338

5072

$\left[\frac{A \times O}{I} = F\right]$ $\left[S \times F \times T = U\right]$ $\left[\frac{P}{U} = C\right]$

Reverse Calculation.
S.S. "Jose Baro."

$$C = 1.12 \overline{\smash{\big)}\, 752} \; (671.4 = U$$
$$\begin{array}{r} \underline{672} \\ 800 \\ \underline{784} \\ 160 \\ \underline{112} \\ 480 \\ \underline{448} \\ \underline{\underline{32}} \end{array}$$

$$T = 302.9°\overline{\smash{\big)}\, 671.4} \;(2.216573 = B$$
$$\begin{array}{r} \underline{605.8} \\ 6560 \\ \underline{6058} \\ 5020 \\ \underline{3029} \\ 19910 \\ \underline{18174} \\ 17360 \\ \underline{15145} \\ 22150 \\ \underline{21203} \\ 9470 \\ \underline{9087} \\ \underline{\underline{383}} \end{array}$$

$$\begin{array}{r} 13.45 = F \\ \underline{1728} = I \end{array}$$

$$O = 24''\overline{\smash{\big)}\, 23241.60} \;(968.4 = A$$
$$\begin{array}{r} \underline{216} \\ 10760 \\ \underline{9415} \\ 1845 \\ \underline{2690} \\ 164 \\ \underline{144} \\ 201 \\ \underline{192} \\ 96 \\ \underline{96} \\ \underline{\underline{}} \end{array}$$

$$S = .1648 \overline{\smash{\big)}\, 2.216573}\;(13.45 = F$$
$$\begin{array}{r} \underline{1648} \\ 5685 \\ \underline{4944} \\ 7417 \\ \underline{6592} \\ 8253 \\ \underline{8240} \\ \underline{\underline{13}} \end{array}$$
lb.

$$\left[\frac{P}{C} = U\right] \quad \left[\frac{U}{T} = B\right] \quad \left[\frac{B}{S} = F\right] \quad \left[\frac{F \times I}{O} = A\right]$$

[89]

Direct Calculation.

S. S. "Normanton." Messrs. Oswald & Co., Sunderland.

```
25" dia. = 490·87  =  area  of  high  pressure
               23 = cut off O    [cylinder = A
          ------
           1462
          98174
          ------
I = 1728)1129·001(6·53 = cubical contents
         10368            of supply in feet = F
         -----
           9220
           8640
          -----
           5801
           5184
          -----
            617
          =====
```

```
      lb.
   ·1314 = weight  of  1 cubic foot of  steam
    6·53 = F                 at 40·3 lbs. above
   -----                     atmosphere = S
    3942
   6570
  7884
  -------
 ·858042 = B
  287·1°= sensible temperature in foot degrees = T
  -------
  858042
 6006294
686 336
1716084
---------
246·3438582 = units of heat = U
```

```
  P = Total indicated H. P.
U = 246·34) 322·25 (1·3 = C
            246·34
            ------
             75910
             73902
            ------
              2008
            ======
```

$$\left[\frac{A \times O}{I} = F\right] \quad \left[S \times F \times T = U\right] \quad \left[\frac{P}{U} = C\right]$$

Reverse Calculation.

S. S. "Normanton."

```
            P
C = 1·3) 322·25 (247·88 = U
         26
         ---
          62
          52
          ---
          102
           91
          ---
           115
           104
           ---
           110
           104
           ---
             6
           ===

  lb.        B
S = ·1314) ·863392 (6·57 = F
            7884
            ----
             7499
             6570
             ----
              9292
              9198
              ----
                94
              ====
```

```
               U
T = 287·1°) 247·88 (·863392 = B
            22968
            -----
             18200
             17226
             -----
               9740
               8613
              -----
               11270
                8613
               -----
                 26570
                 25839
                 -----
                   7310
                   87 2
                   ----
                   1568
                   ====
```

```
             1728 = I
             6·57 = F
            ------
            12096
            8640
           10368
          --------
O = 23") 11352·96 (493·606 = A
          92
          ---
          215
          207
          ---
           82
           69
           ---
           139
           138
           ---
            160
            138
            ---
             22
             ==
```

$$\left[\frac{P}{C} = U\right] \quad \left[\frac{U}{T} = B\right] \quad \left[\frac{B}{S} = F\right] \quad \left[\frac{F \times I}{O} = A\right]$$

DIRECT CALCULATION.

S. S. "SAVERNAKE." MESSRS. OSWALD & CO., SUNDERLAND.

23″ dia. = 415·47 = area of high pressure cylinder = A
15″ = cut of = O

```
              415·47
                  15
            --------
              207735
              41547
            --------
I = 1728 ) 6232·05 ( 3·6 = cubical contents
              5184         of supply in feet = F
            --------
              10480
              10368
            --------
                112
```

```
   lb.
 ·1627 = weight of 1 cubic foot of steam
   3·6 = F     at 54·3 lbs. above atmosphere = S
 --------
   9762
  4881
 --------
 ·58572 = B
  301·9° = sensible temperature in foot degrees = T
 --------
  527148
  58572
 1757160
 --------
 176·828868 = units of heat = U
```

P = total indicated H. P.

```
U = 176·82 ) 332·60 ( 1·88 = C
              17682
            --------
              155780
              141456
            --------
              143240
              141456
            --------
                1784
```

$$\left[\frac{A \times O}{I} = F\right] \quad \left[S \times F \times T = U\right] \quad \left[\frac{P}{U} = C\right]$$

Reverse Calculation.

S. S. "Savernake."

```
             P
O = 1·88) 332·6 (176·9 = U
          188
          ----
          1446
          1316
          ----
           1300
           1128
           ----
            1720
            1692
            ----
              28
            ====
```

```
              U
T = 301·9°) 176·90 (·58595 = B
            15095
            -----
             25950
             24152
             -----
              17980
              15095
              -----
               28850
               27171
               -----
                16790
                15095
                -----
                 1695
                =====
```

```
              1728 I =
               3·6 = F
             -------
             10368
             5184
             -------
O = 15″)   6220·8 (414·7 = A
           60
           --
            22
            15
            --
             70
             60
             --
             108
             105
             ---
               3
             ===
```

```
                 B
     lb.
S = ·1627) ·58595 (3·6 = F
            4881
            -----
             9785
             9762
             ----
               23
             ====
```

$\left[\frac{P}{O} = U\right]$ $\left[\frac{U}{T} = B\right]$ $\left[\frac{B}{S} = F\right]$ $\left[\frac{F \times I}{O} = A\right]$

DIRECT CALCULATION—FULL POWER.

S. S. "WALLACE." MESSRS. OSWALD & CO., SUNDERLAND.

```
45" dia. = 1590·43  =  area of high pressure
              23"   =  cut off O     [cylinder = A
           ------
           477129
          318086
           ------
I = 1728)36579·89(21·16 = cubic contents
          3456             of supply in feet = F
          ------
           2019
           1728
          ------
            2918
            1728
          ------
           11909
           10368
          ------
            1541
          ======

P = Total indicated H. P.
U = 847·99)1030·00(1·21 =
             847·99
            -------
            182010
            169598
            -------
             124120
              84799
            -------
              39321
            =======
```

```
         lb.
      ·1380 = weight of 1 cubic foot of steam
      21·16 = F                  43·3 lbs. above
     ------                    atmosphere = S
       8280
      1380
     1380
    2760
    --------
    2·920080 = B
       290·4° = sensible temperature in foot
                       degrees T =
    --------
     11680320
   262807200
   5840160
   ------------
   847·9912820 = units of heat = U
```

$$\left[\frac{A \times O}{I} = F\right] \quad \left[S \times F \times T = U\right] \quad \left[\frac{P}{U} = C\right]$$

Reverse Calculation.

S. S. "Wallace."

```
         P
C = 1·21)1030(851·23 = U
          968
          ---
           620
           605
           ---
            150
            121
            ---
             290
             242
             ---
              480
              363
              ---
              117

          B
    lb. )2·931232(21·24 = F
S = ·1380)2760
         ----
          1712
          1380
          ----
           3323
           2760
           ----
            5632
            5520
            ----
             112

  T       U
290·4°) 851·28 (2·931232 = B
        5808
        ----
        27043
        26136
        -----
          9070
          8712
          ----
           3580
           2904
           ----
            6760
            5808
            ----
             9520
             8712
             ----
              8080
              5808
              ----
              2272

           21·24 = F
            1728 = I
           -----
           16992
           4248
          14868
          2124
          --------
O = 23")36702·72 (1595·77 = A
        23
        ---
        137
        115
        ---
         220
         207
         ---
          132
          115
          ---
           177
           161
           ---
            162
            161
            ---
              1
```

$$\left[\frac{P}{C} = U\right] \quad \left[\frac{U}{T} = B\right] \quad \left[\frac{B}{S} = F\right] \quad \left[\frac{F \times I}{O} = A\right]$$

Direct Calculation—Half Power.

S. S. "Wallace." Messrs. Oswald & Co., Sunderland.

45″ dia. = 1590·43 = area of high pressure [cylinder = A

6″ = cut off = O

```
I = 1728) 9542·58 (5·52 = cubical contents
          8640          of supply in feet = F
          ----
           9025
           8640
           ----
            3858
            3456
            ----
             402
```

P = total indicated H. P.

```
U = 298·57) 575·000 (1·92 = C
            29857
            ------
            276430
            268713
            ------
              77170
              59714
              -----
              17456
```

```
    lb.
  ·1759 = weight of 1 cubic foot of steam
   5·52 = F          60·3 lbs. above
  -----              atmosphere = S
   3518
   8795
  8795
  -------
·970968 = B
  307·5° = sensible temperature in foot
                  degees = T
  -------
   4854840
  6796776
 29129040
 -----------
298·5726600 = units of heat = U
```

$$\left[\frac{A \times O}{I} = F\right] \quad \left[S \times F \times T = U\right] \quad \left[\frac{P}{U} = C\right]$$

Revelse Calculation.

S. S. "Wallace."

```
            P
C = 1·92) 575 (299·47 = U
          384
          ----
          1910
          1728
          ----
          1820
          1728
          ----
           920
           768
          ----
          1520
          1344
          ----
           176
          ====

   lb.      B
S = ·1759) ·973886 (5·53 = F
            8795
           -----
            9438
            8795
           -----
            6436
            5277
           -----
            1159
           =====
```

```
              U
T=307·5°) 299·47 (·973886 = B
          27675
          -----
          22720
          21525
          -----
          11950
           9225
          -----
          27250
          24600
          -----
          26500
          24600
          -----
          19000
          18450
          -----
            550
          =====
```

```
           1728 = I
           55·3 = F
          -----
           5184
          8640
         8640
         -------
O = 6°) 9555·84 (1592·64 = A
        6
        --
        35
        30
        --
         55
         54
         --
          15
          12
          --
           38
           36
           --
            24
            24
            ==
```

$$\left[\frac{P}{C} = U\right] \quad \left]\frac{U}{T} = B\right] \quad \left[\frac{B}{S} = F\right] \quad \left[\frac{F \times I}{O} = A\right]$$

Direct Calculation.

S. S. "Arndt," Messrs. Oswald & Co., Sunderland.

51″ dia. = 2042·82 = area of high pressure cylinder = A
7″ = cut off = O

I = 1728) 14299·74 (8·27 = cubical contents of supply in feet = F
13824
———
4757
3456
———
13014
12096
———
918

P = Total indicated H. P.

U = 475·4) 1550·0 (3·26 = C
14262
———
12380
9508
———
28720
28524
———
196

lb.
·1848 = weight of 1 cubic foot of steam at 64·3 lbs. above atmosphere = S
8·27 = F
———
12936
3696
14784
———
1·528296 = B
311·1° = sensible temperature in foot degrees = T
———
1528296
1528296
1528296
4584888
———
475·4528856 = units of heat = U

$$\left[\frac{A \times O}{I} = F\right] \quad \left[S \times F \times T = F\right] \quad \left[\frac{P}{U} = C\right]$$

REVERSE CALCULATION.

S. S. "ARNDT."

```
            P                          U                            8·26 = F
C = 3·26) 1550 (475·4 = U   T = 311·1°) 475·4 (1·528126 = B          1728 = I
          1304                          3111                         ----
          ----                          -----                        6608
           2460                         16430                       1652
           2282                         15555                      5782
           ----                         -----                      ·826
            1780                          8750                     --------
            1630                          6222           O = 7") 14273·28 (2039·04 = A
            ----                          -----                   14
             1500                         25280                   --
             1304                         2 888                    27
             ----                         -----                    21
              196                          3920                    --
             ====                          3111                    63
                                           ----                    63
      lb.   B                               8090                   --
S = ·1848) 1·528126 (8·26 = F               6222                    28
           14784                            -----                   28
           -----                            18680                  ====
            4972                            18666
            3696                            -----
            -----                              14
            12766                            ====
            11088
            -----
             1678
             ====
```

$$\left[\frac{P}{C} = U\right] \quad \left[\frac{U}{T} = B\right] \quad \left[\frac{B}{S} = F\right] \quad \left[\frac{F \times I}{O} = A\right]$$

Direct Calculation.

S. S. "Dhoolia," Messrs. Oswald & Co., Sunderland.

45″ dia. = 1590·43 = area of high pressure [cylinder = A.
21″ = cut off = O

```
        159043
       318086
I = 1728)33399 03 (19·32 = cubical contents
        1728            of supply in feet = F.
        16119
        15552
         5670
         5184
          4863
          3456
          1407
```

```
 lb.
 ·1627 = weight of 1 cubic foot of steam
 19·32 = F            at 54·3 lbs. above
                      atmosphere = S
  3254
  4881
 14643
 1627
 3·143364 = B
    301·9° = sensible temperature in foot
                     degrees = T
 28290276
 3143364
94300920
948·9815916 = units of heat = U
```

P = total indicated H. P.

```
U = 948·98) 1097·60 (1·15 = C
            94898
            148620
             94898
             537220
             474490
              62730
```

$$\left[\frac{A \times O}{I} = F\right] \quad \left[S \times F \times T = U\right] \quad \left[\frac{P}{U} = C\right]$$

Reverse Calculation.
S. S. "Dhoolia."

```
              P
C = 1·15)1097·6 (954·43 = U
         1035
         ----
          626
          575
          ---
          510
          460
          ---
          500
          460
          ---
          400
          345
          ---
           55
          ===

   lb.      B
S = ·1627)3·161411 (19·43 = F
          1627
          -----
          15344
          14643
          -----
           7011
           6508
           ----
           5031
           4881
           ----
            150
           ====
```

```
             U
T = 301·9°) 954·43 (3·161411 = B
            9057
            -----
            4873
            3019
            -----
            18540
            18114
            -----
             4260
             3019
             -----
             12410
             12076
             -----
              3340
              3019
              ----
              3210
              3019
              ----
               191
              ====
```

```
          19·43 = F
          1728 = I
          -----
          15544
          3886
         13601
         1943
         --------
O = 21") 33575·04 (1598·81 = A
         ·21
         ---
          125
          105
          ---
          207
          189
          ---
           185
           168
           ---
           170
           168
           ---
            24
            21
            --
             3
            ==
```

$$\left[\frac{C}{P} = U\right] \quad \frac{U}{T} = U\Big] \quad \left[\frac{B}{S} = F\right] \quad \left[\frac{F \times I}{O} = A\right]$$

Direct Calculation.

S. S. "Patroclus," Robert Stephenson and Co., Newcastle-on-Tyne.

28¼" dia. = 626·79 = area of high pressure [cylinder = A
O = cut off = 28"

```
                501432
               125358
              ---------
I = 1728 ) 17550·12 ( 10·15 = cubical contents
               1728            of supply in feet = F
              ---------
                2701
                1728
              ---------
                 9732
                 8640
              ---------
                 1092
```

P = total indicated H. P.

```
U = 591·87 ) 732·00 ( 1·23 = C
             591·87
            ---------
             140130
             118374
            ---------
              217560
              177561
            ---------
               39999
```

```
  lb.
 ·1869 = weight of 1 cubic foot of
  10·15=F    steam at 65·3 above
             atmosphere = S
 -------
   9345
  1869
 18690
 -------
 1·897035 = B
      312° = sensible temperature in foot
                 degrees = T
 -------
  3794070
 1897035
5691105
 -------
591·874920 = units of heat = U
```

$$\left[\frac{A \times O}{I} = F\right] \quad \left[S \times F \times T = U\right] \quad \left[\frac{P}{U} = C\right]$$

Reverse Calculation.

S. S. "Patroclus."

```
                P
C = 1·23) 732 (595·12 = U
          615
          ----
          1170
          1107
          ----
           630
           615
           ----
           150
           123
           ----
            270
            246
            ----
             24
             ====
```

```
              U
T = 312°) 595·12 (1·907435 = B
          312
          ----
          2831
          2808
          ----
           2320
           2184
           ----
            1360
            1248
            ----
             1120
              936
             ----
             1840
             1560
             ----
              280
              ====
```

```
                 10·20 = F
                 1728  = I
                 -----
                  8160
                 2040
                7140
               1020
               --------
O = 28") 17625·60 (629·48 = A
         168
         ----
          82
          56
          ----
          265
          252
          ----
           136
           112
           ----
            240
            224
            ----
             16
             ====
```

```
 lb.           B
S = ·1869) 1·907435 (10·20 = F
           1869
           ----
            3843
            3738
            ----
             1055
             ====
```

$$\left[\frac{P}{C} = U\right] \quad \left[\frac{U}{T} = B\right] \quad \left[\frac{B}{S} = F\right] \quad \left[\frac{F \times I}{O} = A\right]$$

Direct Calculation.

S. S. "Olbers," Messrs. Stephenson & Co. Newcastle-on-Tyne.

29" dia. = 660·52 = area of high pressure [cylinder = A.
26" = cut off = O

```
        396312
       132104
      ────────
I = 1728) 17173·52 (9·93 = cubical contents
          15552           of supply in feet = F
          ─────
           16215
           15552
           ─────
             6632
             5184
             ────
             1448
```

P = Total indicated H. P.

```
U = 638·72) 1062·21 (1·66 = C
             63872
             ──────
             423490
             383232
             ──────
              402580
              383232
              ──────
               19348
```

```
      lb.
     ·2024 = weight of 1 cubic foot of steam
      9·93 = F        at 72·3 lbs. above
                      atmosphere = S
     ─────
      6072
     18216
    18216
    ────────
    2·009832 = B
      317·8° = sensible temperature in foot
                        degrees = T
    ────────
    16078656
   14068824
   2009832
  6029496
  ────────────
  638·7246096 units of heat = U
```

$$\left[\frac{A \times O}{I} = F\right] \quad \left[S \times F \times T = U\right] \quad \left[\frac{P}{U} = C\right]$$

REVERSE CALCULATION.

S. S. "OLBERS."

```
         P
C = 1·66) 1062·21 (639·88 = U
          996
          ---
           662
           498
           ---
           1641
           1494
           ----
            1470
            1328
            ----
             1420
             1328
             ----
               92
```

```
           U
T = 317·8°) 639·88 (2·013467 = B
            6356
            ----
              4280
              3178
              ----
              11020
               9534
              -----
               14860
               12712
               -----
                21480
                19068
                -----
                 24120
                 22246
                 -----
                  1874
```

```
            1728 = I
            9·94 = F
            --------
            6912
           15552
          15552
          ------------
O = 26") 17176·32 (660·62 = A
         156
         ---
          157
          156
          ---
           163
           156
           ---
             72
             52
             --
             20
```

```
  lb.          B
S = ·2024) 2·013467 (9·94 = F
           18216
           -----
            19186
            18216
            -----
              9707
              8096
              ----
              1611
```

$$\left[\frac{P}{C} = U\right] \quad \left[\frac{U}{T} = B\right] \quad \left[\frac{B}{S} = F\right] \quad \left[\frac{F \times I}{O} = A\right]$$

DIRECT CALCULATION.

S. S. "ARISTOCRAT," MESSRS. R. STEPHENSON & CO., NEWCASTLE-ON-TYNE.

24" dia. 452·39 = area of high pressure cylinder = A.
27" = cut off = O

```
     316673
      90478
     --------
I = 1728) 12214·58 (7·06 = cubical contents of supply in feet = F
          12096
          -----
           11853
           10368
           -----
            1485
```

lb.
·1980 = weight of 1 cubic foot of steam at 70·3 lbs. above atmosphere = S
7· = F

1·3860 = B
316·1° = sensible temperature in foot degrees = T

```
     13860
    83160
   13860
  41580
  ---------
  438·11460 units of heat = U
```

P = total indicated H. P.

```
U = 438·11) 645·00 (1·47 = O
            43811
            ------
            206890
            175244
            ------
             316460
             306677
             ------
               9783
```

$$\left[\frac{A \times O}{I} = F\right] \quad \left[S \times F \times T = U\right] \quad \left[\frac{P}{U} = O\right]$$

Reverse Calculation.

S. S. "Aristocrat."

```
          P
C = 1·47) 645 (438·77 = U
          588
          ---
           570
           441
           ---
           1290
           1176
           ----
            1140
            1029
            ----
            1110
            1029
            ----
              81
```

```
               U
T = 316·1°) 438·77 (1·388 = B
            3161
            ----
            12267
             9483
            -----
            27840
            25288
            -----
             25520
             25288
             -----
               232
```

```
          1728 = I
             7 = F
          ---------
O = 27″) 12096 (448· = A
         108
         ---
          129
          108
          ---
          216
          216
```

```
  lb.       B
S = ·1980) 1·3880 (7 = F
           13860
           -----
              20
```

$\left[\frac{P}{C} = U\right]$ $\left[\frac{U}{T} = B\right]$ $\left[\frac{B}{S} = F\right]$ $\left[\frac{F \times I}{O} = A\right]$

CHAPTER X.

FORMULÆ TO OBTAIN THE LOSS OF HEAT IN THE STEAM IN COMPOUND-ENGINE CYLINDERS.

THE proportions that require attention in this case must be arranged in the following order for the "direct" formulæ:—

Indicated horse power collectively.

Speed of piston in feet per minute.

Mean pressure of steam in high pressure cylinder in lbs. per square inches.

Mean pressure of steam in low pressure cylinder in lbs. per square inches.

Area of high pressure cylinder in square inches.

Area of low pressure cylinder in square inches.

Revolutions of crank pin per minute.

The mean pressure in the high pressure cylinder H, added to the mean pressure in the low pressure cylinder L, equals the working mean pressure W.

The area of the high pressure cylinder D, added to the area of the low pressure cylinder E, equals the collective area A.

The collective area A, multiplied by speed of piston S, equals the surface of exertion Y.

Next, indicated horse power P, multiplied by 33000 K, equals motive power V.

The motive power V, divided by the surface of exertion Y, equals the theoretical mean pressure Z.

The working mean pressure W, divided by the theoretical mean pressure Z, equals the steam constant C. We have condensed those formulæ in the form as shown, thus—

$$\left[A \times S = Y\right] \quad \left[P \times K = V\right] \quad \left[\frac{V}{Y} = Z\right]$$

$$\left[\frac{W}{Z} = C\right]$$

The reverse calculation must be arranged thus. The working mean pressure W, divided by the steam constant C, equals the theoretical mean pressure Z.

Next the indicated horse power P, multiplied by K, equals V.

Then the motive power V, divided by the theoretical mean pressure Z, equals the surface of exertion Y.

The surface of exertion Y, divided by the speed of piston S, equals the collective cylinders areas.

We have also condensed those formulæ in the form as shown thus—

$$\left[\frac{W}{C} = Z\right] \left[P \times K = V\right] \left[\frac{V}{Z} = Y\right] \left[\frac{Y}{S} = A\right]$$

Calculations to obtain the Loss of Heat in the Steam in Compound Engines.

Direct Calculation.—S. S. "Mongolia," Messrs. Day, Summers, & Co., Southampton.

Indicated horse power = 1258 = P
Speed of piston in feet per minute = 450 = S
Mean pressure of steam in high pressure cylinder = 22·12 = H lbs.
„ „ Low „ = 7·23 = L

Working mean pressure = 29·35 = W

Z = Theoretical mean pressure. C = constant
Revolutions per minute = 50. Stroke = 54″

Dia. of high pressure cylinder = 48″ area = 1809·56 = D
„ Low „ „ = 96″ „ = 7238·24 = E

Collective area = 9047·80 = A

K = 33000

```
    9047 8 = A          1258 = P
       450 = S         33000 = K
   ---------          ----------
   4523900            3774000
   361912             3774
   ---------          ----------
 4071510·0 = Y ) 41514000 = V (10·19 = Z
                 40715100
                 ---------
                  79890000
                  40715100
                 ----------
                 391749000
                 366435900
                 ----------
                  25313100
```

```
                  W
 Z = 10·19 ) 29·35 (2·88 = C
             2038
             ----
              8970
              8152
             ----
              8180
              8152
             ----
                28
             ====
```

$\left[A \times S = Y\right]$ $\left[P \times K = V\right]$ $\left[\frac{V}{Y} = Z\right]$ $\left[\frac{W}{Z} = C\right]$

Reverse Calculation.

S. S. "Mongolia."

```
           W                        1258 = P                          Y
C = 2·88) 29·35 (10·19 = Z        83000 = K            S = 450) 4073994 (9053·32 = A
          288                   ---------                       4050
          ---                    3774000                        -------
           550                  3774                             2399
           288                  ---------                        2250
          ----       Z = 10·19) 41514000 = V (4073994· = Y       ----
           2620                 4076                              1494
           2592                 -----                             1350
           ----                  7540                             ----
             28                  7133                              1440
           ====                  ----                              1350
                                  4070                             ----
                                  3057                              900
                                  -----                             900
                                  10130                             ===
                                   9171
                                   ----
                                   9590
                                   9171
                                   ----
                                    4190
                                    4076
                                    ----
                                     114
```

$\left[\frac{W}{C} = Z\right]$ $\left[P \times K = V\right]$ $\left[\frac{V}{Z} = Y\right]$ $\left[\frac{Y}{S} = A\right]$

111]

Direct Calculation.

S. S. "Danube," Messrs. Day, Summers & Co., Southampton.

Indicated horse power = 862·3 = P　　Revolutions per minute = 54.　Stroke = 50″

Speed of piston in feet per minute = 449·82 = S

Mean pressure of steam in high pressure cylinder = 34·38 lbs. = L

" " Low " " = 6·95 = L

Working mean pressure = 41·33 = W

Dia. of high pressure cylinder = 36″ area = 1017·87 = D

" Low " " = 72″ " = 4071·51 = E

Collective area = 5089·38 = A

```
Z = 12·42) 41·33 (3·3 = C
           3726
           ----
           4070
           3726
           ----
            344
```

```
   5089·38 = A
    449·82 = S
  ---------
   1017876
  4071504
 4580442
2035752
2035752
------------
2289204·9116 = Y)28455900·000 = V(12·42 = Z
```

```
862·3 = P
  33000 = K
---------
25869000
25869
```

$\left[A \times S = Y\right]$ $\left[P \times K = V\right]$ $\left[\frac{V}{Y} = Z\right]$ $\left[\frac{W}{Z} = C\right]$

Reverse Calculation.

C = 3·3) 41·33 (12·52 = Z　(W)

```
862·3 = P
  33000 = K
---------
25869000
25869
```

S = 449·82) 2272835·46 (5052 = A　(Y)

$\left[\frac{W}{C} = Z\right]$ $\left[P \times K = V\right]$ Z = 12·52)28455900·0(2272835·46 = Y $\left[\frac{V}{Z} = Y\right]$ $\left[\frac{Y}{S} = A\right]$

Direct Calculation.

S. S. "Lady Josyan," Messrs. Day, Summers & Co., Southampton.

Indicated horse power = 294 = P
Speed of piston in feet per minute = 288 = S
Mean pressure of steam in high pressure cylinder = 26·3 = H lbs.
" " low " " = 9·3 = L

Working mean pressure = 35·6 = W

Z = theoretical mean pressure. C = constant
Revolutions per minute = 48. Stroke = 36″

```
                W
Z = 12·68) 35·60 (2·8 = C
           25·36
           -----
           10240
           10144
           -----
              96
```

Dia. of high pressure cylinder = 26″ area = 530·93 = D
" low " " = 52″ " = 2123·72 = E

Collective area = 2654·65 = A

```
2654·65 = A
    288 = S        294 = P
                 33000 = K
 -------         -------
 2123720          882000
2123720           882
530930
-----------      ---------
764539·20 = Y) 9702000·0 = V (12·68 = Z
```

[A × S = Y] [P × K = V] $\left[\frac{V}{Y} = Z\right]$ $\left[\frac{W}{Z} = C\right]$

Reverse Calculation.

```
          W
C = 2·8) 35·6 (12·71 = Z
```

$\left[\frac{W}{C} = Z\right]$ [P × K = V]

```
      294 = P
    33000 = K
    ---------
    882000
    882
    ---------
Z = 12·71) 9702000·0 = V (763335·9 = Y
```

```
               Y
S = 288) 763335·9 (2650·4 = A
```

$\left[\frac{V}{Z} = Y\right]$ $\left[\frac{Y}{S} = A\right]$

Direct Calculation.

S. S. "Amerique," Messrs. Maudslay, Sons & Field, London.

Revolutions per minute = 60. Stroke = 51"

Indicated horse power = 1617 = P
Speed of piston in feet per minute = 514·25 = S
Mean pressure of steam in high pressure cylinder = 39 lbs. = H
" " low " " = 11·85 = L

Working mean pressure = 50·85 = W

Dia. of high pressure cylinder = 41" area = 1320 = D
" low " " = 75" " = 4417 = E

Collective area = 5737 = A

$$Z = 18{\cdot}08\,\big)\;\overset{W}{50{\cdot}85}\;\big(\,2{\cdot}8 = C$$

```
 3616
 -----
 14690
 14464
 -----
   226
```

```
   5737 = A
 514·25 = S
 ---------
   28685
  11474
 22948
 5737
28685
---------------
2950252·25 = Y) 53361000·0 = V (18·08 = Z
```

```
 1617 = P
33000 = K
---------
4851000
4851
```

$$\left[A \times S = Y\right] \quad \left[P \times K = V\right] \quad \left[\frac{V}{Y} = Z\right] \quad \left[\frac{W}{Z} = C\right]$$

Reverse Calculation.

$$C = 2{\cdot}8\,\big)\;\overset{W}{50{\cdot}85}\;\big(\,18{\cdot}16 = Z$$

$$\left[\frac{W}{C} = Z\right] \quad \left[P \times K = V\right]$$

```
 1617 = P
33000 = K
---------
4851000
4851
---------
Z = 18·16) 53361000 = V (2938381·05 = Y
```

$$S = 514{\cdot}25\,\big)\;\overset{Y}{2938381{\cdot}05}\;\big(\,5713 = A$$

$$\left[\frac{V}{Z} = Y\right] \quad \left[\frac{Y}{S} = A\right]$$

DIRECT CALCULATION.

S. S. "TIMOR," MESSRS. MAUDSLAY, SONS & FIELD, LONDON.

Indicated horse power = 1234 = P   Revolutions per minute = 65.   Stroke = 45″

Speed of piston in feet per minute = 495 = S

	lbs.
Mean pressure of steam in high pressure cylinder	= 44·00 = H
" " low " "	= 10·35 = L
Working mean pressure	= 54·35 = W

Dia. of high pressure cylinder = 36″ area = 1017 = D

" low " " = 68″ " = 3631 = E

Collective area = 4648 = A

```
         W
Z = 17·6) 54·35 (3·08 = C
          528
          ----
          1550
          1408
          ----
           142
```

```
 4648 = A
  495 = S        1234 = P
 -----             33000 = K
 23240           ---------
41832            3702000
18592            3702
---------        ---------
2300760 = Y) 40722000 = V ( 17·6 = Z
```

$\left[A \times S = Y\right]$ $\left[P \times K = V\right]$ $\left[\frac{V}{Y} = Z\right]$ $\left[\frac{W}{Z} = C\right]$

REVERSE CALCULATION.

```
          W
C = 3·08) 54·35 (17·6 = Z
```

$\left[\frac{W}{C} = Z\right]$ $\left[P \times K = V\right]$

```
1234 = P
  33000 = K
---------
3702000
3702
---------
Z = 17·6) 40722000 = V (2313750 = Y
```

```
           Y
S = 495) 2313750 (4674 = A
```

$\left[\frac{V}{Z} = Y\right]$ $\left[\frac{Y}{S} = A\right]$

Direct Calculation.

S. S. "Nankin," Messrs. Maudslay, Sons & Field, London.

Revolutions per minute = 59. Stroke = 48"

Indicated horse power = 1221 = P
Speed of piston in feet per minute = 472 = S
Mean pressure of steam in high pressure cylinder = 42·6 = H lbs.
" " low " " = 10·1 = L

Working mean pressure = 52·7 = W

Dia. of high pressure cylinder = 38" area = 1134 = D
" low " " = 70" " = 3848 = E

Collective area = 4982 = A

```
Z = 17·13) 52·70 (3·07 = C
           5139
           -----
           13100
           11991
           -----
            1109
```

```
   4982 = A         122 = P
    472 = S          33000 = K
  -------          -----------
   9964            3663000
  34874            3663
 19928
------------------------------
2351504 = Y) 40293000 = V (17·13 = Z
```

[A × S = Y] [P × K = V] [$\frac{V}{Y}$ = Z] [$\frac{W}{Z}$ = C]

Reverse Calculation.

```
C = 3·07) 52·7 (17·16 = Z
             W
```

[$\frac{W}{C}$ = Z] [P × K = V]

```
       1221 = P
        33000 = K
     -----------
     3663000
     3663
     -----------
Z = 17·16) 40293000 = V (2348076· = Y
```

```
S = 472) 2348076 (4974 = A
            Y
```

[$\frac{V}{Z}$ = Y] [$\frac{Y}{S}$ = A]

Direct Calculation.

S. S. "Peter Jebson," Messrs. Maudslay, Sons & Field London.

Indicated horse power = 619 = P

Revolutions per minute = 72. Stroke = 30"

Speed of piston in feet per minute = 360 = S

Mean pressure of steam in high pressure cylinder = 40·8 = H lbs.

" " low " " = 12·15 = L

Working mean pressure = 52·95 = W

Dia. of High pressure cylinder = 29" area = 660 = D

" low " " = 56" " = 2463 = E

Collective area = 3123 = A

```
                 W
Z = 18·16) 52·95 (2·9 = C
           3632
           -----
           16630
           16344
           -----
             286
```

```
 3123 = A        619 = P
  360 = S      33000 = K
 -------       -------
 187380        1857000
 9369          1857
 -------       --------
1124280 = Y) 20427000 = V (18·16 = Z
```

$\left[A \times S = Y\right]$ $\left[P \times K = V\right]$ $\left[\frac{V}{Y} = Z\right]$ $\left[\frac{W}{Z} = C\right]$

Reverse Calculation.

```
             W
C = 2·9) 52·95 (18·25 = Z
```

$\left[\frac{W}{C} = Z\right]$ $\left[P \times K = V\right]$

```
                 619 = P
               33000 = K
               -------
               1857000
               1857
               --------
Z = 18·25)  20427000 = V (1119287·67 = Y
```

```
                  Y
S = 360) 1119287·67 (3109·13 = A
```

$\left[\frac{V}{Z} = Y\right]$ $\left[\frac{Y}{S} = A\right]$

Direct Calculation.

S. S. "Garonne," Messrs. N. Napier & Sons, Glasgow.

Indicated horse power = 2650 = P — Revolutions per minute = 62. Stroke = 48″

Speed of piston in feet per minute = 488 = S

Mean pressure of steam in high pressure cylinder = 22·1 lbs. = H

" " low " " = 13·80 = L

Working mean pressure = 35 90 = W

Dia. of high pressure cylinder = 60″ area = 2827·44 = D

" low " " = 104″ " = 8494·88 = E

Collective area = 11322·32 = A

```
 11322·32 = A
      488 = S        2650 = P
                      33000 = K
  9057856
 9057856             7950000
4528928              7950

5525292·16 = Y ) 87450000·0 = V ( 15·82 = Z
```

```
              W
Z = 15·82 ) 35·90 ( 2·26 = C
             3164
             4260
             3164
            10960
             9492
             1468
```

$\left[A \times S = Y\right]$ $\left[P \times K = V\right]$ $\left[\frac{V}{Y} = Z\right]$ $\left[\frac{W}{Z} = C\right]$

Reverse Calculation.

```
            W
C = 2·26 ) 35·90 ( 15·88 = Z
```

$\left[\frac{W}{C} = Z\right]$ $\left[P \times K = V\right]$

```
 2650 = P
  33000 = K

 7950000
7950

Z = 15·88 ) 87450000·0 = V ( 5506926·95 = Y
```

```
                 Y
S = 488 ) 5506926·95 ( 11284 68 = A
```

$\left[\frac{V}{Z} = Y\right]$ $\left[\frac{Y}{S} = A\right]$

DIRECT CALCULATION.

S. S. "E. M. ARNDT," MESSRS. OSWALD & CO, SUNDERLAND.

Indicated horse power = 1550 = P Revolutions per minute = 60. Stroke = 42″

Speed of piston in feet per minute = 420 = S

Mean pressure of steam in high pressure cylinder = 39 lbs. = H

" " low " " = 10·45 = L

Working mean pressure = 49·45 = W

Dia. of high pressure cylinder = 51″ area = 2042·82 = D

" low " " 86″ " = 5808·81 = E

Collective area = 7851·63 = A

```
W
Z = 15·51) 49·45 (3·18 = G
           46 53
           -----
            2920
            1551
           -----
           13690
           12408
           -----
            1282
```

```
7851·63 = A          1550 = P
    420 = S         33000 = K
-----------        ---------
 15703260            4650000
3140652             4650
-----------        ---------
3297684·60 = Y) 51150000·0 = V( 15·51 = Z
```

$\left[A \times S = Y\right]$ $\left[P \times K = V\right]$ $\left[\frac{V}{Y} = Z\right]$ $\left[\frac{W}{Z} = C\right]$

REVERSE CALCULATION.

```
             W
C = 3·18) 49·45 (15·55 = Z
```

$\left[\frac{W}{C} = Z\right]$ $\left[P \times K = V\right]$

```
             1550 = P
            33000 = K
           ---------
            4650000
           4650
           ---------
Z = 15·55) 51150000 = V (3289389·06 = Y
```

$\left[\frac{V}{Z} = Y\right]$

```
                 Y
S = 420) 3289389·06 (7831·87 = A
```

$\left[\frac{Y}{S} = A\right]$

Direct Calculation—Full Power.

S. S. "Wallace," Messrs. Oswald & Co., Sunderland.

Indicated horse power = 1030 = P   Revolutions per minute = 52.   Stroke = 42″

Speed of piston in feet per minute = 364 = S

		lbs.	
Mean pressure of steam in high pressure cylinder	=	31·12	= H
" " low " "	=	8·00	= L
Working mean pressure	=	39·12	= W

Dia. of high pressure cylinder = 45″ area = 1590·43 = D

" low " " = 82″ " = 5281·00 = E

Collective Area = 6871·43 = A

$$Z = 13{\cdot}58\ \big)\ \overset{W}{39{\cdot}12}\ \big(\ 2{\cdot}88 = C$$

```
   39·12
   2716
   -----
   11960
   10864
   -----
   10960
   10864
   -----
      96
```

```
6871·43 = A
    364 = S
-----------
 2748572
4122858
2061429
-----------
2501200·52 = Y
```

```
1030 = P
  33000 = K
-----------
 3090000
3090
-----------
33990000·0 = V
```

$$2501200{\cdot}52 = Y\ \big)\ 33990000{\cdot}0 = V\ \big(\ 13{\cdot}58 = Z$$

$$\left[A \times S = Y\right] \quad \left[P \times K = V\right] \quad \left[\frac{V}{Y} = Z\right] \quad \left[\frac{W}{Z} = C\right]$$

$$C = 2{\cdot}88\ \big)\ \overset{W}{39{\cdot}12}\ \big(\ 13{\cdot}58 = Z$$

$$\left[\frac{W}{C} = Z\right] \quad \left[P \times K = V\right]$$

Reverse Calculation.

```
1030 = P
  33000 = K
-----------
 3090000
3090
```

$$Z = 13{\cdot}58\ \big)\ 33990000{\cdot}0 = V\ \big(\ 2502945{\cdot}5 = Y$$

$$S = 364\ \big)\ \overset{Y}{2502945{\cdot}5}\ \big(\ 6876{\cdot}2 = A$$

$$\left[\frac{V}{Z} = Y\right] \quad \left[\frac{Y}{S} = A\right]$$

Direct Calculation—Half Power.

S. S. "Wallace," Messrs. Oswald & Co., Sunderland.

Indicated horse power = 575 = P

Revolutions per minute = 40. Stroke = 42"

Speed of piston in feet per minute = 280 = S

Mean pressure of steam in high pressure cylinder = 26·7 lbs. = H

„ „ low „ „ = 5·4 = L

Working mean pressure = 32·1 = W

Dia. of high pressure cylinder = 45" area = 1590·43 = D

„ low „ „ = 82" „ = 5281·00 = E

Collective area = 6871·43 = A

```
            W
Z = 9·86) 32·10 ( 3·25 = C
          2958
          ----
           2520
           1972
           ----
            5480
            4930
            ----
             550
```

```
 6871·43 = A        575 = P
     280 = S      33000 = K
 -----------      -----------
  54971440        1725000
 1374286          1725
 -----------      -----------
1924000·40 = Y) 18975000·00 = V (9·86 = Z
```

$\left[A \times S = Y\right]$ $\left[P \times K = V\right]$ $\left[\frac{V}{Y} = Z\right]$ $\left[\frac{W}{Z} = C\right]$

Reverse Calculation.

```
              W
C = 3·25) 32·10 (9·87 = Z
```

$\left[\frac{W}{C} = Z\right]$ $\left[P \times K = V\right]$

```
       575 = P
     33000 = K
     ---------
     1725000
     1725
     ---------
Z = 9·87) 18975000 (1922492·4 = Y
```

```
            Y
S = 280) 1922492·4 (6866 0 = A
```

$\left[\frac{V}{Z} = Y\right]$ $\left[\frac{Y}{S} = A\right]$

DIRECT CALCULATION.

S. S. "JOSE BARO," MESSRS. OSWALD & CO., SUNDERLAND.

Indicated horse power = 752 = P. Revolutions per minute = 74. Stroke = 36"

Speed of piston in feet per minute = 444 = S

Mean pressure of steam in high pressure cylinder = 37·2 = H lbs.

" " low " " = 5·9 = L

Working mean pressure = 43·1 = W

Dia. of high pressure cylinder = 35" area = 962·11 = D

3421·20 = E

Collective area = 4383·31 = A

$$Z = 12{\cdot}7 \overset{W}{\big)\ 43{\cdot}1\ \big(}\ 3{\cdot}39 = C$$

```
   381
   ---
   500
   381
  ----
  1190
  1143
  ----
    47
```

```
  4383·31 = A         753 = P
      444 = S       33000 = K
 ------------      ---------
  1753324           2256000
 1753324           2256
1753324
------------       ------------
1946189·64 = Y) 24816000·0 = V ( 12·7 = Z
```

$$\left[A \times S = Y\right] \quad \left[P \times K = V\right] \quad \left[\frac{V}{Y} = Z\right] \quad \left[\frac{W}{Z} = C\right]$$

REVERSE CALCULATION.

$$C = 3{\cdot}39 \overset{W}{\big)\ 43{\cdot}1\ \big(}\ 12{\cdot}17 = Z$$

$$\left[\frac{W}{C} = Z\right] \quad \left[P \times K = V\right]$$

```
      752 = P
    33000 = K
   ---------
   2256000
  2256
  ---------------
Z = 12·7) 24816000·0 = V ( 1954015·7 = Y
```

$$S = 444 \overset{Y}{\big)\ 1954015{\cdot}7\ \big(}\ 4400{\cdot}9 = A$$

$$\left[\frac{V}{Z} = Y\right] \quad \left[\frac{Y}{S} = A\right]$$

DIRECT CALCULATION.

S. S. "SAVERNAKE," MESSRS. OSWALD & Co., SUNDERLAND.

Indicated horse power = 332 6 = P Revolutions per minute = 78. Stroke = 30″

Speed of piston in feet per minute = 390 = S

Mean pressure of steam in high pressure cylinder = 37 7 lbs. = H

" " low " " 9·15 = L

Working mean pressure = 46·85 = W

Dia. of high pressure cylinder = 23″ area = 415·47 = D

" low " " = 42″ " = 1385·44 = E

Collective area = 1800·91 = A

$$Z = 15{\cdot}6\,\big)\ \overset{W}{46{\cdot}85}\ \big(\,3{\cdot}003 = C$$

```
 46·85
 468
 ----
  500
  468
 ----
   32
```

```
  1800·91 = A        332·6 = P
      390 = S        33000 = K
  ---------        ---------
  16208190          9978000
  540273           9978
  ---------        ---------
  702354·90 = Y ) 10975800·0 = V ( 15·6 = Z
```

$$\left[A \times S = Y\right] \quad \left[P \times K = V\right] \quad \left[\frac{V}{Y} = Z\right] \quad \left[\frac{W}{Z} = C\right]$$

REVERSE CALCULATION.

$$C = 3\,\big)\ \overset{W}{46{\cdot}85}\ \big(\,15{\cdot}60 = Z$$

$$\left[\frac{W}{C} = Z\right] \quad \left[P \times K = V\right]$$

```
   332·6 = P
   33000 = K
  ---------
  9978000
  9978
  ---------
Z = 15·6 ) 10975800·0 = V ( 703576·92 = Y
```

$$S = 390\,\big)\ \overset{Y}{703576{\cdot}92}\ \big(\,1804{\cdot}04 = A$$

$$\left[\frac{V}{Z} = Y\right] \quad \left[\frac{Y}{S} = A\right]$$

DIRECT CALCULATION.

S. S. "NORMANTON," MESSRS. OSWALD & Co., SUNDERLAND.

Indicated horse power = 322·25 = P — Revolutions per minute = 63. Stroke = 33"

Speed of piston in feet per minute = 346·5 = S

Mean pressure of steam in high pressure cylinder = 35·275 lbs. = H

" " low " " = 7·38 = L

Working mean pressure = 42·655 = W

Dia. of high pressure cylinder = 25" area = 490·87 = D

" low " " = 48" " = 1809·56 = E

Collective area = 2300·43 = A

$$Z = 13{\cdot}34 \overset{W}{\big)} \; 42{\cdot}655 \; \big(3{\cdot}19 = C$$

```
 W
42 655
4002
------
 2635
 1334
------
13010
12006
------
 1004
```

```
 2300·43 = A
   346·5 = S
 ---------
 1150215
1380258
920172
690129
-----------
797098·995 = Y
```

```
322·25 = P
  33000 = K
 ----------
 96675000
96675
```

$$797098{\cdot}995 = Y \big) \; 10634250{\cdot}00 = V \; \big(13{\cdot}34 = Z$$

$$\left[A \times S = Y \right] \quad \left[P \times K = V \right] \quad \left[\frac{V}{Y} = Z \right] \quad \left[\frac{W}{Z} = C \right]$$

REVERSE CALCULATION.

$$C = 3{\cdot}39 \overset{W}{\big)} \; 42{\cdot}655 \; \big(13{\cdot}37 = Z$$

$$\left[\frac{W}{C} = Z \right] \quad \left[P \times K = V \right]$$

```
322·25 = P
  33000 = K
 ----------
 96675000
96675
```

$$Z = 13{\cdot}37 \big) \; 10634250{\cdot}00 = V \; \big(795381\ 4 = Y$$

$$S = 346{\cdot}5 \overset{Y}{\big)} \; 795381{\cdot}4 \; \big(2295{\cdot}4 = A$$

$$\left[\frac{V}{Z} = Y \right] \quad \left[\frac{Y}{S} = A \right]$$

DIRECT CALCULATION.

S. S. "DHOOLIA," MESSRS. OSWALD & CO. SUNDERLAND.

Indicated horse power = 1097·6 = P
Speed of piston in feet per minute = 364 = S
Mean pressure of steam in high pressure cylinder = 38 lbs. = H
" " low " " = 10 = L

Working mean pressure = 48 = W

Dia. of high pressure cylinder = 45″ area = 1590·43 = D
" low " " = 82″ area = 5281·02 = E

Collective area = 6871·45 = A

Revolutions per minute = 52. Stroke = 42″

$$Z = 14{\cdot}48\,)\ \overset{W}{48{\cdot}00}\ (\,3{\cdot}31 = C$$

```
       W
Z = 14·48) 48·00 (3·31 = C
           43 48
           -----
            4520
            4348
            ----
             1720
             1448
             ----
              272
```

```
 6871·45 = A
     364 = S      1097·6 = P
 -------            33000 = K
 2748580          ----------
4122870            32928000
2061435           32928
----------        ----------
2501207·80 = Y) 36220800·0 = V ( 14·48 = Z
```

$$\left[A \times S = Y\right] \quad \left[P \times K = V\right] \quad \left[\frac{V}{Y} = Z\right] \quad \left[\frac{W}{Z} = C\right]$$

REVERSE CALCULATION.

```
            W
C = 3·31) 4·80 (14·5 = Z
```

```
        1097·6 = P
          33000 = K
        ----------
         32928000
        32928
        ----------
Z = 14·5) 36220800·0 = V (2497986·2 = Y
```

```
             Y
S = 364) 2497986·2 (6862·5 = A
```

$$\left[\frac{W}{C} = Z\right] \quad \left[P \times K = V\right] \quad \left[\frac{V}{Z} = Y\right] \quad \left[\frac{Y}{S} = A\right]$$

DIRECT CALCULATION.

S. S. "OLBERS," MESSRS. ROBERT STEPHENSON & CO., NEWCASTLE-ON-TYNE.

Indicated horse power = 1062 = P
Revolutions per minute = 56 75 Stroke = 48″
Speed of piston in feet per minute = 454 = S
Mean pressure of steam in high-pressure cylinder = 62·2 = H
" " low " " = 10·55 = L

Working mean pressure = 72·75 = W

Dia. of high pressure cylinder = 29″ area = 660·5 = D
" low " " = 66″ " = 3421·2 = E

Collective area = 4081·7 = A

```
              W
Z = 18·91) 72·75 (3·84 = C
           56 73
           -----
           16020
           15128
           -----
            8920
            7564
           -----
            1356
```

```
 4081·7 = A        1062 = P
    454 = S          33000 = K
 ---------        ---------
 163268            3186000
204085             3186
163268
-----------       ---------
1853091·8 = Y) 35046000 = V ( 18·91 = Z
```

$\left[A \times S = Y\right]$ $\left[P \times K = V\right]$ $\left[\frac{V}{Y} = Z\right]$ $\left[\frac{W}{Z} = C\right]$

REVERSE CALCULATION.

```
          W
C = 3·84) 72·75 (18 94 = Z
```

```
  1062 = P
   33000 = K
  ---------
  3186000
 3186
 ---------
Z = 18·94) 35046000 = V (1850369·5 = Y
```

```
          Y
S = 454) 1850369·5 (4075·7 = A
```

$\left[\frac{W}{C} = Z\right]$ $\left[P \times K = V\right]$ $\left[\frac{V}{Z} = Y\right]$ $\left[\frac{Y}{S} = A\right]$

Direct Calculation.

S. S. "Patroclus," Messrs. Stephenson & Co., Newcastle-on-Tyne.

Indicated horse power = 732 = P — Revolutions per minute = 52. Stroke = 45″

Speed of piston in feet per minute = 390 = S

Mean pressure of steam in high pressure cylinder = 50·35 = H

" " low " " = 10·35 = L

Working mean pressure = 60·70 = W

Dia. of high pressure cylinder = 28½″ area = 626·79 = D

" low " " = 61″ " = 2922·47 = E

Collective area = 3549·26 = A

```
        W
Z = 17·4) 60·70 (3·4 = C
          522
          ---
          850
          696
          ---
          154
```

```
3549·26 = A            732 = P
    390 = S          33000 = K
-----------          ---------
 31943340             2196000
1064778              2196
-----------          ---------
1384211·40 = Y) 24156000·0 = V (17·4 = Z
```

$\left[A \times S = Y\right]$ $\left[P \times K = V\right]$ $\left[\dfrac{V}{Y} = Z\right]$ $\left[\dfrac{W}{Z} = C\right]$

Reverse Calculation.

```
          W
C = 3·4) 60·70 (17·8 = Z
```

$\left[\dfrac{W}{C} = Z\right]$ $\left[P \times K = V\right]$

```
              732 = P
            33000 = K
            ---------
            2196000
           2196
            ---------
Z = 17·8) 24156000 = V (1357078·5 = Y
```

```
              Y
S = 390) 1357078·5 (3479·6 = A
```

$\left[\dfrac{V}{Z} = Y\right]$ $\left[\dfrac{Y}{S} = A\right]$

DIRECT CALCULATION.

S. S. "ARISTOCRAT," MESSRS. ROBERT STEPHENSON & CO., NEWCASTLE-ON-TYNE.

Indicated horse power = 645 = P
Speed of piston in feet per minute = 432 = S
Mean pressure of steam in high pressure cylinder = 61·75 = H
" " low " " = 11·9 = L

Working mean pressure = 73·65 = W

Dia. of high pressure cylinder = 24" area = 452·39 = D
" low " " = 48" " = 1809·56 = E

Collective area = 2261·95 = A

Revolutions per minute = 72. Stroke = 36"

```
Z = 21·78) 73·65  (3·38 = C
           6534
           ----
            8310
            6534
            ----
            17760
            17424
            -----
              336
```

```
 2261·95 = A          645 = P
     432 = S        33000 = K
 -------          --------
  452390          1935000
 678585           1935
904780            -----------
---------
977162·40 = Y) 21285000·0 = V (21·78 = Z
```

$\left[A \times S = Y\right]$ $\left[P \times K = V\right]$ $\left[\frac{V}{Y} = Z\right]$ $\left[\frac{W}{Z} = C\right]$

REVERSE CALCULATION.

```
              W
C = 3·38) 73·65 (21·78 = Z
```

```
        645 = P
      33000 = K
   ---------
    1935000
   1935
   -----------
Z = 21·78) 21285000·0 = V (977272·72 = Y
```

```
              Y
S = 432) 977272·72 (2262·2 = A
```

$\left[\frac{W}{C} = Z\right]$ $\left[P \times K = V\right]$ $\left[\frac{V}{Z} = Y\right]$ $\left[\frac{Y}{S} = A\right]$

CHAPTER XI.

MEMORANDA, RULES, AND TABLES.

Pressure of Steam for Compound-Engines.—The main consideration in this case is the amount of expansion agreed on. Our advice is, Expand twenty times, if possible; as, for example: We should use the initial steam, at 200lbs. on the square inch, and we should exhaust into the condenser at 10lbs. the square inch. We ignore a vacuum becuuse it is worthless in comparison to feed water entering the boiler at 208 deg. Fahr., in the place of 90 deg., because $208-90=118$; then, as that sum represents 118 deg. of heat, it represents also, at 200lbs. on the square inch pressure, more than one-third of that temperature.

There is another great fact, also, by the absence of the vacuum, and that is, the temperature of the low-pressure cylinder is not cooled down to the extent it would be if the vacuum were caused.

The cooling of the cylinders, whether by exhaust steam, liquefaction, or radiation, is really the great difficulty to be overcome, and it is for that reason that a great number of engineers have preferred to consider the temperature of the exhaust steam in proportion to the indicated horse power obtained, their reason being that the most power they could get from the steam would be represented by the exhausting pressure in the cylinder.

In the most modern practice the average initial pressure for the steam is about 70 lbs. on the square inch, while the average mean pressure in the high pressure cylinder is about 30, and the average mean pressure in the low pressure cylinder is about 10.

Returning back again to the matter of the pressure of the steam at the point of exhaustion, it requires explanation that, that pressure is independent of the condenser, and therefore is due to expansion only.

Now here comes a great fact to be noticed. Theoretically the pressure of the steam at the point of exhaustion, as shown by a diagram, would be due to the "finishing off" of the hyperbolical expansion curve.

But we have explained on page 59 that in practice that curve is more full than it should be, consequently the pressure at the point of exhaustion is shown by the indicator diagram to be much more than it really is. We know this as a positive fact by proving the matter practically from a long stroke beam engine, working at about ten strokes per minute; length of stroke twelve feet, the cylinder being steam jacketed, surrounded by earth, wood, and felt casing enclosed with brickwork; length of cut-off, one-twelfth of the piston's stroke. An almost perfect expansion curve was recorded by the indicator, because there was but very little loss by radiation, and time was allowed for the expansion of the steam, as also for the instrument to record it.

Now, with a quick-working compound engine the steam has time allowed to expand, but not enough time is allowed for the diagram to be taken, and the result is, as we have said before, the diagram is erroneous, but on the right side to show more duty than performed and on the wrong side to show a truthful recordance. The result is that the theory of the hyperbolical curve and the hyperbolical logarithm said to belong to it is much distorted.

We, however, have no objection to the hyperbolical logarithm provided sufficient allowance is admitted for the loss in the elastic force of the steam due to radiation and other equally important matters which we have explained.

It is for those reasons, there
and steam constants, C, have been introduced for the first time, as shown in Chapters IX and X. In fact, why we inserted so many calculations is to prove the veracity of our statements as taken from practical results.

To Find the Mean Pressure of the Steam in Lbs. per Square Inch in the Cylinders of a Compound Engine.—Multiply the length of the piston stroke after the cut off in the high pressure cylinder in inches *a* by the area of that cylinder in square inches *b* equals *e*.

Multiply the stroke of the piston *d* by ·8 equals *i*.

Multiply the area of the low pressure cylinder in square inches *c* by the resultant *i* equals *f*. Then *f* added to *e* equals *g*. Next *g* divided by *b* equals *h*, and *h* plus E, or length of cut off, equals S, or elongated stroke.

The elongated stroke S divided by E equals the ratio F. Then the initial pressure in pounds per square inch I divided by F equals the multiplier G.

Next the hyperbolic logarithm of the ratio F equals the expansion grade logarithm H, and H plus 1 equals the multiplicand T, which multiplied by G equals the calculated mean pressure K.

The calculated mean pressure K divided by the steam constant C equals the theoretical mean pressure Z.—

$$\left[a \times b = e\right]\left[d \times \cdot 8 = i\right]\left[c \times i = f\right]\left[f + e = g\right]$$

$$\left[\frac{g}{b} = h\right] \quad \left[h + \mathrm{E} = \mathrm{S}\right] \quad \left[\frac{\mathrm{S}}{\mathrm{E}} = \mathrm{F}\right] \quad \left[\frac{\mathrm{I}}{\mathrm{F}} = \mathrm{G}\right]$$

$$\left[\mathrm{H} + 1 = \mathrm{T}\right] \quad \left[\mathrm{T} \times \mathrm{G} = \mathrm{K}\right] \quad \left[\frac{\mathrm{K}}{\mathrm{C}} = \mathrm{Z}\right]$$

Mean Pressure Calculations for Both Cylinders.
S. S. "Garonne," Messrs. R. Napier & Sons, Glasgow.

a = 24″ remainder of stroke after cut off
b = 2827·44 area of high pressure cylinder
c = 8494·88 area of low pressure cylinder
d = 48″ stroke in inches
$e = a \times b$
$f = c \times i$
$g = e + f$
$h = \frac{g}{b}$

C = Steam constant
E = Length of cut off
F = Ratio
G = Multiplier
H = Expansion grade logarithm
I = Initial pressure
K = Calculated mean pressure
S = Elongated stroke
T = Multiplicand
Z = Mean pressure theoretically

```
 2827·44 = b
     24″ = a
 -------
 1130976
 565488
 -------
67858·56 = e
```

```
                8494·88 = c
d = 48 × ·8 =     38·4 = i
                -------
                 339
                6795904
               2548464
               -------
               326     = f
                67858·56 = e
               --------
```

```
139·37 = h
   24· = E
------
163·37 = S
```

$E = 24″ \,)\, 163·37 = S \,(\, 6·80 = F$

$b = 2827·44 \,)\, 394061·952 = g \,(\, 139·37 = h$

lbs.

$F = 6·80 \,)\, 44·3 = I \,(\, 6·51 = G$

Mean pressures as indicated
High pressure cylinder = 22·10
Low pressure cylinder = 13·80

```
2) 35·90
   -----
Mean = 17·95
```

Hyperbolic logarithm of F = 6·80 = 1·91692 = H

```
   1·
-------
2·91692 = T
   6·51 = G
-------
 291692
1458460
1750152
----------
18·9891492 = K
```

$C = 2·27 \,)\, 18·9891492 = K \,(\, 8·36526 = Z$

Mean Pressure Calculations for Both Cylinders.

S. S. "Danube."

$a = 24''$ after cut off
$b = 1017{\cdot}87$ square inches
$c = 4071{\cdot}51$ square inches
$d = 50''$ piston stroke
$E = 26''$ cut off

$$\begin{array}{r} 1017{\cdot}87 = b \\ 24'' = a \\ \hline 407148 \\ 203574 \\ \hline 24428{\cdot}88 = e \end{array}$$

$$\begin{array}{r} 4071{\cdot}51 = c \\ d = 50'' \times {\cdot}8 = 40 = i \\ \hline 162860{\cdot}40 = f \\ 24428{\cdot}88 = e \\ \hline \end{array}$$

$$b = 1017{\cdot}87 \Big) \; 187289{\cdot}28 = g \; \Big(184{\cdot}00 = h$$

$$\begin{array}{r} 184{\cdot}00 = h \\ 26{\cdot}00 = E \\ \hline 210 = s \end{array}$$

$S = 210''$
$E = 26''$
$I = 50{\cdot}3$ lbs.
$H =$ hyperbolic logarithm of F
$T = H \times I$

$$E = 26'' \Big) \; 210 = s \; \Big(8{\cdot}07 = F$$

$$F = 8{\cdot}07 \Big) \; 50{\cdot}3 \text{ lbs.} = I \; \Big(6{\cdot}23 = G$$

Hyperbolic logarithm of $F = 8{\cdot}07 = 2{\cdot}08815 = H$

$$\begin{array}{r} 2{\cdot}08815 = H \\ 1 \\ \hline 3{\cdot}08815 = T \\ 6{\cdot}23 = G \\ \hline 926445 \\ 617630 \\ 1852890 \\ \hline 19{\cdot}2391745 = Z \end{array}$$

$$C = 3{\cdot}3 \Big) \; 19{\cdot}2391745 = K \; \Big(5{\cdot}830052 = Z$$

Mean pressures as indicated
High pressure cylinder = 34·38
Low pressure cylinder = 6·95

$$2 \Big) \; 41{\cdot}33$$

Mean = 20·66

Mean Pressure Calculations for Both Cylinders.

S. S. "Lady Josyan."

a = 18″ after cut off
b = 530·93 square inches
c = 2123·72 square inches
d = 36″ piston stroke
E = 18″ cut off

$$\begin{array}{r} 530{\cdot}93 = b \\ 18'' = a \\ \hline 424744 \\ 53093 \\ \hline 9556{\cdot}74 = e \\ \hline \end{array}$$

$$d = 36'' \times {\cdot}8 = \begin{array}{r} 2123{\cdot}72 = c \\ 28{\cdot}8'' = i \\ \hline 1698976 \\ 1698976 \\ 424744 \\ \hline 61163{\cdot}136 = f \\ 9556{\cdot}74 = e \\ \hline \end{array}$$

$$\begin{array}{r} 133{\cdot}2 = h \\ 18{\cdot}0 = E \\ \hline 151{\cdot}2 = S \\ \hline \end{array}$$

$$b = 530{\cdot}93 \,\big)\, 70719{\cdot}876 = g \,\big(\, 133{\cdot}2 = h$$

S = 151·2″
E = 18″
I = 42·3 lbs.
H = hyperbolic logarithm of F
T = H × 1

$$E = 18'' \,\big)\, 151{\cdot}2 = s \,\big(\, 8{\cdot}4 = F \qquad F = 8{\cdot}4 \,\big)\, 42{\cdot}3 \text{ lbs.} = I \,\big(\, 5{\cdot}03 = G$$

Hyperbolic logarithm of F = 8·4 = 2·12823 = H

$$\begin{array}{r} 2{\cdot}12823 = H \\ 1{\cdot} \\ \hline 3{\cdot}12823 = T \\ 5{\cdot}03 = G \\ \hline 938469 \\ 15641150 \\ \hline 15{\cdot}7349969 = Z \end{array}$$

Mean pressures as indicated
High pressure cylinder = 26·30
Low pressure cylinder = 9·30

$$2 \,\big)\, \begin{array}{r} 35{\cdot}60 \\ \hline \text{Mean } 17{\cdot}80 \end{array} \qquad C = 2{\cdot}8 \,\big)\, 15{\cdot}7349969 = K \,\big(\, 5{\cdot}619641 = K$$

TABLE OF HYPERBOLIC LOGARITHM.

No.	Hyperbolic Logarithm.	No.	Hyperbolic Logarithm.	No.	Hyperbolic Logarithm	No.	Hyperbolic Logarithm.
1	·0000	3·75	1·32175	6·5	1·87180	11	2·39790
1·25	·22314	4	1·38629	6·75	1·90954	12	2·48491
1·5	·40546	4·25	1·44691	7	1·94591	13	2·56495
1·75	·55961	4·5	1·50507	7·25	1·98100	14	2·63906
2	·69315	4·75	1·55814	7·5	2·01490	15	2·70805
2·25	·81093	5	1·60944	7·75	2·04769	16	2·77259
2·5	·91629	5·25	1·65822	8	2·07944	17	2·83321
2·75	1·01160	5·5	1·70474	8·5	2·14006	18	2·89037
3	1·09861	5·75	1·74919	9	2·19722	19	2·94444
3·25	1·17865	6	1·79176	9·5	2·25129	20	2·99573
3·5	1·25276	6·25	1·83258	10	2·30259	21	3·04452

TO FIND THE MEAN PRESSURE OF THE STEAM IN POUNDS PER SQUARE INCH IN THE HIGH PRESSURE CYLINDER OF A COMPOUND ENGINE.—Divide the length of the piston stroke in inches, S, by the length of the cut off in inches E equal F.

Divide the initial pressure I by F equals G.

To the hyperbolic logarithm H of the sum of F add 1 equals T.

Multiply T by G equals the theoretical mean pressure of the steam Z.

Then Z divided by the steam constant C × ·5 equals the actual mean pressure N.

$$\left[\frac{S}{E} = F\right]\left[\frac{I}{F} = G\right]\left[H + 1 = T\right]\left[T \times G = Z\right]$$

$$\left[\frac{Z}{C} = N\right]$$

The following calculations indicate the practical utility of the formulæ—

Mean Pressure Calculations for High Pressure Cylinder.

S. S. "Garonne."

= 24") 48" = S

F = 2) 44·3 lbs. = I

22·15 = G

Logarithm of 2 = F = ·69315 = H

```
   1·0
 1·69315 = T
   22·15 = G
  846575
 169315
338630
338630
```

C = 2·26 × ·5 = 1·13) 37·5032725 = Z (33·18873 = N

$\left[\frac{S}{E} = F\right]$ $\left[\frac{I}{F} = G\right]$ $\left[H + 1 = T\right]$ $\left[T \times G = Z\right]$ $\left[\frac{Z}{C} = N\right]$

S. S. "Danube."

E = 26") 50" = S

F = 1·92) 50·3 lbs. = I

26·19 = G

Logarithm of 1·92 = F = ·65233 = H

```
   1·0
 1·65233 = T
   26·19 = G
 1487097
 165233
991398
330466
```

C = 3·32 × ·5 = 1·66) 43·2745227 = Z (26·06898 = N

$\left[\frac{S}{E} = F\right]$ $\left[\frac{I}{F} = G\right]$ $\left[H + 1 = T\right]$ $\left[T \times G = Z\right]$ $\left[\frac{Z}{C} = N\right]$

Mean Pressure Calculations for High Pressure Cylinder.

S. S. "Lady Josyan."

E = 18″) 36″ = S

F = 2) 42·3 lbs. = I

21·15 = G

Logarithm of 2 F = ·69315 = H
·1·0

1·69315 = T
21·15 = G

846575
169315
169315
338630

C = 2·8 × ·5 = 1·4) 35·8101225 = Z (25·57865 = N

$$\left[\frac{S}{E} = F\right] \quad \left[\frac{I}{F} = G\right] \quad \left[H + 1 = T\right] \quad \left[T \times G = Z\right] \quad \left[\frac{Z}{C} = N\right]$$

To find the Pressure of Steam at the Point of Exhaustion from an Engine Cylinder separately.—Multiply the length of the piston's stroke by ·8 = S, divide that sum by the length of stroke allowed for the initial steam to enter the cylinder or "length of cut off," E equals the ratio F. Then the initial pressure in lbs. per square inch I divided by the ratio F equals the pressure at the point of exhaustion T. Therefore the following formula:—

$$\left[\frac{S}{F} = F\right] \quad \left[\frac{1}{F} = T\right]$$

To find the Pressure of the Steam at the Point of Exhaustion from the Low Pressure Cylinder of a Compound Engine.—Multiply the area of the high pressure cylinder in square inches *a* by the length of cut-off in inches *b* = A.

Multiply the remainder of the piston stroke from the cut off *c* by the area of the same cylinder *a* = B.

Multiply the length of the piston stroke *d* by ·8, and that result by the area of the low pressure cylinder *e* = C.

Add the sums of B and C together *f* and divide that result by A = D.

Initial steam pressure in lbs. per square inch = I.

Then I divided by D equals the pressure of steam at the point of exhaustion theoretically = E.

"Exhaust" Steam Calculations.

S. S. "Garonne," Messrs. Napier & Sons, Glasgow.

a = 2827·44 square inches
b = 24 inches
c = 24 inches
d = 48 inches
e = 8494·88 square inches
I = 44·3 lbs.
E = pressure of steam at point of exhaustion theoretically

square inches
2827·44 = a
24″ = b

1130976
565488

67858·56 = A

square inches
2827·44 = a
24″ = c

1130976
565488

67858·56 = B

48″ = d
·8

38·4 = i

square inches
8494·88 = e
38·4 = i

3397952
6795904
2548464

326203·392 = C
67858·56 = B

A = 67858·56) 394061·952 = f (5·80 = D

lbs.
D 5·80) 44·3 = I (7·63 = E

$[a \times b = A]$ $[a \times c = B]$ $[d \times \cdot 8 = i]$ $[i \times e = C]$ $[C + B = F]$ $\left[\frac{f}{A} = D\right]$ $\left[\frac{I}{D} = E\right]$

"Exhaust" Steam Calculations.

S. S. "Danube," Messrs. Day, Summers & Co., Southampton.

$a = 1017{\cdot}87$ square inches
$b = 26$ inches
$c = 24$ inches
$d = 50$ inches
$e = 4071{\cdot}51$ square inches
$I = 50{\cdot}3$ lbs.
E = pressure of steam at point of exhaustion theoretically

$$\begin{array}{r} \text{square inches} \\ 1017{\cdot}87 = a \\ 26'' = b \\ \hline 610722 \\ 203574 \\ \hline 26464{\cdot}62 = A \end{array} \qquad \begin{array}{r} \text{square inches} \\ 1017{\cdot}87 = a \\ 24'' = c \\ \hline 407148 \\ 203574 \\ \hline 24428{\cdot}88 = B \end{array} \qquad \begin{array}{r} 50'' = d \\ {\cdot}8 \\ \hline 40{\cdot}0 = i \end{array}$$

$$\begin{array}{r} \text{square inches} \\ 4071{\cdot}51 = e \\ 40 = i \\ \hline 162860{\cdot}40 = C \\ 24428{\cdot}88 = B \\ \hline \end{array}$$

$$A = 26464{\cdot}62 \,)\, 187289{\cdot}28 = f \,(\, 7{\cdot}07 = D$$

$$D = 7{\cdot}07 \,)\, 50{\cdot}3 \text{ lbs.} = I \,(\, 7{\cdot}11 = E$$

$$\left[a \times b = A\right] \left[a \times c = B\right] \left[d \times {\cdot}8 = i\right] \left[i \times e = C\right] \left[C + B = f\right] \left[\frac{f}{A} = D\right] \left[\frac{I}{D} = E\right]$$

"Exhaust" Steam Calculations.

S. S. "Lady Josyan," Messrs. Day, Summers & Co., Southampton.

a = 530·93 square inches
b = 18 inches
c = 18 inches
d = 36 inches
e = 2123·72 square inches
I = 42·3 lbs.
E = pressure of steam at point of exhaustion theoretically

square inches
530·93 = a
18″ = b

424744
53093

9556·74 = A

square inches
530·93 = a
18″ = c

424744
53093

9556·74 = B

36″ = d
·8

28·8 = i

square inches
2123·72 = e
28·8 = i

1698976
1698976
424744

61163·136 = C
9556·74 = B

A = 9556·74) 70719·876 = f (7·4 = D

lbs.
D = 7·4) 42·3 = I (5·71 = E

$$\left[a \times b = \text{A}\right] \left[a \times c = \text{B}\right] \left[d \times \cdot 8 = i\right] \left[i \times e = \text{C}\right] \left[\text{C} + \text{B} = f\right] \left[\frac{f}{\text{A}} = \text{D}\right] \left[\frac{\text{I}}{\text{D}} = \text{E}\right]$$

SCIENTIFIC TABLE OF THE DUTY EVOLVED BY THE CUBICAL CONTENTS OF THE INITIAL STEAM IN COMPOUND-ENGINE CYLINDERS.

Name of Maker of the Engines.	Units of Heat to equal 1 Indicated Horse Power.	Initial Pressure of Steam.	Cubical Contents of Initial Steam in feet per stroke.	Units of Heat in the Steam per Stroke.	Number of Strokes per Minute.	Indicated Horse Power from both Cylinders.	Working mean Pressure of Steam collectively used.	Theoretical Mean Pressure of Steam.	Steam Constant of Loss of Heat.
Messrs. Day, Summers and Co.	88·66	42·3	28·27	1115·548	100	1258	29·35	10·19	2·88
” ” ”	87·82	50·3	15·3	701·235	108	862·3	41·33	12·42	3·3
” ” ”	71·25	42·3	5·53	218·216	96	294	35·6	12·68	2·8
Messrs. Maudslay, Sons	66·81	54·3	18·33	900·353	120	1617	50·85	18·08	2·8
” ” ”	75·82	62·3	12·9	719·790	130	1234	54·35	17·6	3·08
” ” ”	88·47	65·3	15·7	915·510	11·8	1221	52·7	17·13	3·07
” ” ”	69·47	58·3	5·7	298·662	144	619	52·95	18·16	2·9
Messrs. R. Napier & Sons	75·64	44·3	39·27	1616·593	124	2650	39·90	15·82	2·26
” Oswald & Co.	36·80	64·3	8·27	475·452	120	1550	49·45	15·51	3·18
” ” ”	89·84	54·3	19·32	948·159	104	1097·6	48	14·48	3·31
Full Power.	85·62	43·3	21·16	847·991	104	1030	39·12	13·58	2·88
” ”	131·25	55·3	13·36	1666·903	148	752	43·1	12·7	3·39
Half Power.	41·53	60·3	5·52	298·557	80	575	32·1	9·86	3·25
” ”	82·93	54·3	3·6	176·828	156	332·6	46·85	15·6	3·003
	96·32	40·3	6·53	246·343	·126	322·25	42·655	13·34	3·19
Messrs. Stephenson & Co.	68·26	72·3	9·93	638·724	113·5	1062	72·75	18·91	3·84
” ” ”	84·09	65·3	10·15	591·874	104	732	60·70	17·4	3·4
” ” ”	97·81	70·3	7·06	438·114	144	645	73·65	21·78	3·38

Analysis of the Units of Heat in the Initial Steam in comparison to the Total or Compound Steam Power that produced the Indicated Horse-Power.—What we have to consider now is that there are a certain number of units of heat that produce a stroke of the piston, and that those units of heat are in direct comparison with the mean pressures in the high and low pressure cylinders, and also that those pressures are in direct comparison with the separate areas of the cylinders.

The speed of the piston, it must be remembered, will much depend on the condition of the working surfaces of the engine, and also to the lubricants applied. The proportion of the connecting rod to the crank will affect the pressure on the guide block surface. The "nip" of the piston rod or trunk gland, will also reduce the speed; in fact, the friction of the working parts and the amount of weight in motion are the only resistants to be overcome which when accomplished the speed is the remains of the power issuing.

It must be understood, also, that the mean pressure is a resultant from the proportion of the units of heat to the entire duty it performs; we have, therefore, thought it best to put this matter into formula, as follows:—

FORMULA TO OBTAIN THE COMPOUND STEAM POWER IN THE HIGH AND LOW PRESSURE CYLINDERS.—Multiply the area of the cylinders, high H, or low L, by the mean pressures of the steam, D and E separately, equals the steam power respectively, *a* and *b*, then let

Mean pressure in high pressure cylinder H.
Mean pressure in low pressure cylinder L.
Area of high pressure cylinder D.
Area of low pressure cylinder E.
Compound steam power equals *c*.

$$\left[H \times D = a\right] \quad \left[L \times E = b\right] \quad a + b = c$$

FORMULA TO OBTAIN THE UNIT POWER CONSTANT IN CONNECTION WITH THE UNITS OF HEAT PER STROKE.—Units of heat per stroke U.

Unit power constant = *d*.

$$\frac{c}{U} = d.$$

EXAMPLES OF FORMULÆ—

Compound Steam Power Calculations.—S. S. "Garonne."

Indicated horse power = 2650

Area D = 2827·44
22·1 = H

282744
565488
565488

62486·424 = a

Area E = 8494·88
13·8 = L

6795904
2548464
849488

117229·344 = b

Units of heat per stroke of piston = 1616·598 = U

a = 62486·424
b = 117229·344

179715·768 = c

c

U = 1616·593) 179715·768 (111·16 = d

S. S. "Timor."

Indicated horse power = 1234

Area D = 1017
44 = H

4068
4068

44748 = a

Area E = 3631
10·35 = L

18155
10893
36310

37580·85 = b

Units of heat per stroke of piston = 719·79 = U

a = 44748
b = 37580·85

82328·85 = c

c

U = 719·79) 82328·85 (114·37 = d

S. S. "Nankin."

Indicated horse power = 1221

Area D = 1134
42·6 = H

6804
2268
4536

48308·4 = a

Area E = 3848
10·1 = L

3848
38480

38864·8 = b

Units of heat per stroke of piston = 915·51 = U

a = 48308·4
b = 38864·8

87173·2 = c

c

U = 915·51) 87173·2 (95·21 = d

Indicated horse power = 645

Area D = 452·39
61·75 = H

226195
316673
45239
271434

27935·0825 = *a*

Area E = 1809·56
11·9 = L

1628604
180956
180956

21533 764 = *b*

Units of heat per stroke of piston = 438·11 = U

a = 27935·0825
b = 21533·7640

49468·8465 = *c*

U = 438·11) 49468·8465 (112·91 = *d*

S. S. "Normanton."

Indicated horse power = 322·25

Area D = 490·87
35·275 = H

245435
343609
98174
245435
147261

17315·43925 = *a*

Area E = 1809·56
7·38 = L

1447648
542868
1266692

13354·5528 = *b*

Units of heat per stroke of piston = 246·343 = U

a = 17315·43925
b = 13354·55280

30669·99205 = *c*

U = 246·343) 30669·99205 (124·50 = *d*

S. S. "Lady Josyan."

Indicated horse power = 294

Area D = 530 93
26·3 = H

159279
318558
106186

13963·459 = *a*

Area E = 2123·72
9·3 = L

637116
1911348

19750·596 = *b*

Units of neat per stroke of piston = 218·216 = U

a = 13963·459
b = 19750·596

33714·055 = *c*

U = 218·216) 33714·055 (154·49 = *d*

ANALYSIS OF THE PRINCIPLES THAT GOVERN THE SPEED OF THE PISTON.—The foundation of this subject rests on the fact that the speed of the piston is merely the residue or remains of the power from the steam pressure that remains after all the waste of energy is expended.

We make use of the term "waste of energy,"

power," and equally because, those two resultants when combined absorb the whole of the effective duty of the steam.

When we remember that the areas of the cylinders are multiplied by the mean pressure, we have to consider from whence that mean pressure comes.

Now, as we have before explained, the initial supply steam is really what makes the power of the engine, and therefore the mean pressure is nothing more than the initial power "spread out."

Then, if the initial steam is the basis of the power, we must note the source from which the power is derived. We have explained that in Chapter I., and merely revert to it again as a fact that the speed of the piston, plus loss of heat and friction, is entirely dependent on the units of heat in the steam.

It will be remembered that we have given the

formulæ on page 146 to obtain the unit power constant d, and we now show its application.

FORMULA TO OBTAIN SPEED OF PISTON FROM UNITS OF HEAT IN THE STEAM.—

d = unit power constant.

U = units of heat per stroke of piston.

$$d \times U = c = \text{Compound Steam Power}$$

Indicated Horse Power = P $\quad \left[P \times K = V \right]$

33000 = K $\quad \left[\frac{V}{c} = S \right]$ = Speed of piston in feet per minute.

We have given the following examples on pages 151 and 152, to show the working of this rule, but have omitted the condensed formulæ because it is so simple that it scarcely, if at all, needs repeating.

S. S. "GARONNE."

$d = 111{\cdot}16$
$U = 1616{\cdot}593$
$P = 2650$

```
                 1616·593 = U
                  111·16 = d
              ____________
              9699558     2650 = P
             1616593        33000 = K
            1616593       __________
           1616593        7950000
          1616593         7950
         ___________      __________
c = 179700·47788 ) 87450000·000 = V ( 486·64 = S
```

S. S. "TIMOR."

$d = 114{\cdot}37$
$U = 719{\cdot}790$
$P = 1234$

```
                 719·790 = U
                  114·37 = d
              ____________
              5038530     1234 = P
             2159370        33000 = K
            2879160       __________
            719790        3702000
           719790         3702
         ___________      __________
c = 82322·38230 ) 40722000·000 = V ( 494·66 = S
```

S. S. "NANKIN."

$d = 95{\cdot}21$
$U = 915{\cdot}510$
$P = 1221$

```
                 915·510 = U
                   95·21 = d
              ____________  1221 = P
               915510         33000 = K
             1831020      __________
            4577550       3663000
           8239590        3663
         ___________      __________
c = 87165·70710 ) 40293000·000 = V ( 461·11 = S
```

Piston Speed Calculations.

S. S. "Aristocrat."

$d = 112{\cdot}91$
$U = 438{\cdot}114$
$P = 645$

```
         438·1146 = U
          112·91  = d
       ----------
        4381146        645 = P
      39420314       33000 = K
      8762292  .    ---------
     4381146        1935000
    4381146        1935
    -----------    -----------
c = 49467·519486 ) 21285000 0000 = V ( 430·28 = S
```

S. S. "Normanton."

$d = 124{\cdot}50$
$U = 246{\cdot}343$
$P = 322{\cdot}25$

```
         246·343 = U
           124·5 = d
       ---------        322·25 = P
        1231715          33000 = K
        985372        ----------
       492686        96675000
      246343         96675
      ---------      -----------
c = 30669·7035 ) 10634250·00 = V ( 346·73 = S
```

S. S. "Lady Josyan."

$d = 154{\cdot}49$
$U = 218{\cdot}216$
$P = 294$

```
         218·216 = U
          154·49 = d
       ----------
        1963944      294 = P
        872864     33000 = K
       872864     --------
      1091080      882000
      218216       882
      -----------  -----------
c = 33712·18984 ) 9702000·000 = V ( 287·78 = S
```

153 Scientific Table of the Piston Speed Constants.

Indicated Horse Power.	Units of Heat per stroke.	Speed of Piston in ft. per Minute.	Speed Constants.
2650	1616·593	488·00	3·312
1617	900·353	514·25	1·750
1550	475·452	420·00	1·132
1258	1115·548	450·00	2·478
1234	719·790	495·00	1 454
1221	915·510	472·00	1·939
1097	948·981	364·00	2·607
1062	638·724	454·00	1·406
1030	847·991	364·00	2 329
862	701·235	449·82	1·558
752	666·903	444·00	1·502
732	591 874	390·00	1·517
645	438·114	432·00	1·014
619 } half power	298·662	360·00	0·829
575 } half power	298·557	280·00	1·066
332 } half power	176·828	390·00	0·453
322	246·343	346·50	0·710
294	218·216	288·00	0·757

Speed Constant = Units of Heat per Stroke divided by the Speed of Piston per Minute.

Going back again to Chapter IX. as a reference, we must direct attention to the constant value C.

Now, if that constant is of any good at all, it is applicable to find the speed of the piston, as the units of heat are. as for example:—

FORMULA TO OBTAIN SPEED OF PISTON FROM THE STEAM CONSTANT VALUE.—Divide the actual mean pressure by the constant value C equals the theoretical mean pressure Z, as explained before.

P = Indicated horse power.
Z = Theoretical mean pressure.
Y = Maximum power.
A = Collective area.
V = Power in foot pounds.
K = 33000.
S = Speed of piston in feet per minute.

$$\left[P \times K = V\right] \qquad \left[\frac{V}{Z} = Y\right] \qquad \left[\frac{Y}{A} = S\right]$$

We have given the following examples as the preceding pages 151 and 152, to show the corresponding results from the two basis of the formulæ, which will agree better if more decimals are used.

S. S. "GARONNE."

P = 2650
Z = 15·8
A = 11322·326
K = 33000

```
            2650 = P
           33000 = K
         ---------
         7950000
        7950
         ---------
Z = 15·8) 87450000 = V (5534810·12 = Y
```

A = 11322·326) 5534810·12 = S (488·84 = S

S. S. "TIMOR."

P = 1234
Z = 17·6
A = 4649·567
K = 33000

```
            1234 = P
           33000 = K
         ---------
         3702000
        3702
         ---------
Z = 17·6) 40722000 = V (2313750·00 = Y
```

A = 4649·567) 2313750·00 = Y (497·62 = S

S. S. "NANKIN."

P = 1221
Z = 17·13
A = 4982·577
K = 33000

```
            1221 = P
           33000 = K
         ---------
         3663000
        3663
         ---------
Z = 17·13) 40293000 = V (2352189·14 = Y
```

A = 4982·577) 2352189·14 = Y (472·08 = S

S. S. "ARISTOCRAT."

P = 645
Z = 21·78
A = 2261·951
K = 33000

$$
\begin{array}{r}
645 = P \\
33000 = K \\
\hline
1935000 \\
1935 \\
\hline
\end{array}
$$

$Z = 21{\cdot}78\,\big)\; 21285000 = V\;\big(\,981864{\cdot}09 = Y$

$A = 2261{\cdot}951\,\big)\; 981664{\cdot}09 = Y\;\big(\,434{\cdot}07 = S$

S. S. "NORMANTON."

P = 322·25
Z = 13·34
A = 2300·436
K = 33000

$$
\begin{array}{r}
322{\cdot}25 = P \\
33000 = K \\
\hline
96675000 \\
96675 \\
\hline
\end{array}
$$

$Z = 13{\cdot}34\,\big)\; 10634250{\cdot}00 = V\;\big(\,797170{\cdot}1649 = Y$

$A = 2300{\cdot}436\,\big)\; 797170{\cdot}1649 = Y\;\big(\,346{\cdot}53 = S$

S. S. "LADY JOSYAN."

P = 294
Z = 12·68
A = 2654·651
K = 33000

$$
\begin{array}{r}
294 = P \\
33000 = K \\
\hline
882000 \\
882 \\
\hline
\end{array}
$$

$Z = 12{\cdot}68\,\big)\; 9702000 = V\;\big(\,765141{\cdot}95 = Y$

$A = 2654{\cdot}651\,\big)\; 765141{\cdot}95 = Y\;\big(\,288{\cdot}22 = S$

PROPORTION OF CYLINDERS FOR COMPOUND-ENGINES.—The length of the stroke of the pistons being generally the same in each cylinder, only the areas of the cylinders are often considered.

When two cylinders are used, and the steam cutting off in the high pressure cylinder at about one-third to one-half of the piston's stroke, and the low pressure cylinder having a 12lb. vacuum in it, when insisted on, the area of the high pressure cylinder being 1, the area of the low pressure is 3·25 to 4, or 3·45 may be taken as the average, the initial steam pressure being from 60 to 80lbs.

But beside the areas, the main consideration, after all, is the capacity or cubical contents for the steam to operate in.

Now, let us suppose an area of 1,500 square inches for the high pressure cylinder, length of cut-off 20 inches, and the amount for expansion allowed in that cylinder to be 20 inches more, we then have 60,000 cubic inches of steam ready for the low pressure cylinder, the area of which is say 2·666 times the area of the high pressure cylinder, from which proportion the area 3999 is produced.

We next have to suppose that the length of the limit allowed for further expansion is 40 inches in the low pressure cylinder showing therefrom 159960, which divided by 60·000 equals 2·666; this constant, therefore, comes back in our calculation again subject to the circumstances of proportions.

Now the main question before us is, after all, the amount of duty that can be "got out" of the steam, and that depends much on the proportions allowed for its expansion. Valuable and reliable experiments have been carried out to arrive at the best proportion, but when a certain fact has been arrived at, it has been swept away by the surrounding circumstances that are always altering, so that a large margin on that account is always admitted. The cause for this is the excessive loss of heat in the steam during its working, as fully shown in chapter 10.

The better proportion for two cylinder compound-engines with initial steam at from 80 to 100lbs. on the square inch is cubical contents of initial steam multiplied by eight, which will give the areas of cylinders in the proportion off four to one when the greatest grade of cut-off in the high pressure cylinder is one-half.

Area of Steam Opening caused by the Valve.—The size of this opening in square inches depends entirely on the speed and movement of the valve in combination with the motion of the piston, it being remembered, as we said in

Chapter II., that the "crank pin governs the action of the steam," and we now add that the excentric governs the action of the slide valve.

Now, then, if motion is governed by time, consequently the travel of the expansion valve should allow the initial steam to enter "full" and "cut off" suddenly, and it is for that reason that multiple ports have been introduced for the admission of the initial steam. To the present time we have proved that the travel of the expansion valve should be about one-tenth of the travel of the piston and the travel of the main exhaust valve to be three-fourths to one-half of the expansion valve. We admit, of course, that those fractions are only approximations; but as they are taken from the practice of our best compound-engines they can be relied on.

The proportion of the area of the steam openings again demand our explanation. Of course, strictly speaking, the pressure of the initial steam and area of the high pressure cylinder in proportion to the time allowed for the admission of the steam is the basis; but, then, as the indicated horse-power is deduced from the areas of the cylinders and the mean pressure of the steam—not including what was lost from the latter—we are safer in considering practically the proportion of the areas of the supply openings to the respective areas of the cylinders.

The area of the exhaust opening caused by the valve, is generally twice the area of the supply opening, and it is for that reason that the width of the cylinder supply ports are twice the width of the opening caused by the valve.

The area of the central or main exhaust port should be three times the area of the supply opening caused by the valve.

In arranging the steam ports, great attention must be given to the widths of the bars, or solid portions between the ports; for constructive purposes, those bars should be as narrow as possible.

To enable this matter to be understood, and of utility at the same time, we have given the following table of the proportions which result in the "divisors" or constants to obtain the area of the steam opening in proportion to the area of its cylinder:—

TABLE OF THE PROPORTIONS OF THE STEAM-OPENINGS CAUSED BY THE VALVES IN COMPOUND-ENGINES.

Indicated Power Collectively.	Area of High Pressure Cylinder in square inches.	Area of Steam Opening in square inches.	Divisor for High-Pressure Cylinder.	Area of Low-Pressure Cylinder in square inches.	Area of Steam Opening in square inches.	Divisor for Low-Pressure Cylinder.	MAKERS' NAME.
3500	1809·561	96	18·849	5541·782	228	24·306	Messrs. Gourlay & Co., Dundee.
3000	1320·257	104	12·694	4778·373	299	15·981	,, Maudslay, Sons & Field.
2500	1520·534	52·50	28·962	4071·513	99	41·126	,, R. Napier & Sons, Glasgow.
2334	907·922	75	12·105	4417·875	250	17·671	,, Maudslay, Sons & Field.
2000	1017·878	83·75	12·153	4071·513	180	22·619	,, Day, Summers & Co. Southampton.

We note from those examples that the average divisor for the high pressure cylinder is 16·952, and that the average divisor for the low pressure cylinder is 24·340. The divisors of the areas of the cylinders are thus:—

TABLE OF THE PROPORTIONS OF HIGH AND LOW-PRESSURE CYLINDERS AS PER DIVISOR.

Indicated Power Collectively.	Area of High-Pressure Cylinder in square inches.	Area of Low-Pressure Cylinder in square inches.	Divisor or Constant.
3500	1809·561	5541·782	3·062
3000	1320·257	4778·373	3·619
2500	1520·534	4071·513	2·677
2334	907·922	4417·875	4·865
2000	1017·878	4071·513	4·000

The average divisor is 3·644, from which we conclude that in five compound-engines ranging from 2,000 to 3,500 indicated horse-power the area of pressure high is as 1 to 3·644 for the low pressure cylinder.

The area of steam opening of the high pressure cylinder is one seventeenth and that the low pressure cylinder one twenty-fourth of the area of the respective cylinder.

We give those examples from practice, as a guarantee of our previous remarks being properly based.

Comparative Table of the Proportions of Cylinders of Compound Engines, in connection with the indicated horse power.

Name of Ship.	High Pressure Cylinder.			Low Pressure Cylinder.			
	Indicated horse power.	Length of Stroke in inches.	Area in square inches of the Cyr's. Diameter.	Area in square inches of the Cyr's. Diameter.	Length of Stroke in inches.	Indicated horse power.	Area of Low pressure Cylinder divided by area of high pressure Cylinder.
S. S. 'Mongolia'	545·82	54	1809·56	7238·24	54	712·18	4.00
S. S. 'Danube'	477·00	50	1017·87	4071·51	50	385·30	4.00
S.S. 'Lady Josyan'	121·86	36	530·93	2123·72	36	172·14	4.00
S. S. 'Amerique'	802·23	51	1320·00	4417·00	51	814·77	3.34
S. S. 'Timor'	671·22	45	1017·00	3631.00	45	562·78	3.57
S. S. 'Nankin'	690·95	48	1134·00	3848.00	48	530·05	3.39
S.S. 'Peter Jebson'	293·76	30	660·00	2463·00	30	325·24	3.73
S. S. Garonne'	924·04	48	2827·44	8494·88	48	1725.96	3.00
S. S. 'E. M. Arndt'	1013·98	42	2042·82	5808·81	42	536·02	2 84
S. S. 'Dhoolia'	666·63	42	1590·43	5281·02	42	430·97	3.32
S. S. 'Wallace'*	545·93	42	1590·43	5281·00	42	484·07	3.32
S. S. 'Jose Baro'	581·54	36	962·11	3421·20	36	270·46	3.55
S. S. 'Wallace'	360·30	42	1590·43	5281·00	42	214.70	3.32
S. S. 'Savernake'	185·11	30	415·47	1385·44	30	137.49	3.33
S. S. 'Normanton'	181·81	33	490·87	1809·56	33	140.44	3.68
S. S. 'Olbers'	565·20	48	660·50	3421·20	48	497.20	5.17
S. S. 'Patroclus'	372·96	45	626·79	2922·47	45	359.04	4.66
S. S. 'Aristocrat'	365·69	36	452·39	1809·56	36	279.31	4.00

* Half-power.

PERMANENT LOAD.—We were the first to explain that this "load" affects the speed of the piston to a very great extent. We use the word "permanent" because it may be considered as a constant weight set in motion, the momentum of which is alternately being checked by the operation of the steam at each end of the stroke of the piston.

We use the word "load" because it is a burden imposed upon the engine, although forming a portion of it.

The subjects of the load are the weights of the slide valves, steam pistons, piston rods, guide-blocks, cross-head pins, main connecting rods, air and circulating pumps—pistons and rods, feeding and bilge pump plungers, and, in fact, all details set in motion by the steam pistons.

RULE FOR PERMANENT LOAD.—Add together the weights of the details mentioned in pounds; multiply that sum by the stroke of the piston in feet equals the load at rest.

MOMENTUM LOAD.—This load is the permanent load set in motion, therefore divide the permanent load by the number of strokes per minute, equals the momentum or reciprocating load.

In calculating the areas for the different rods

we prefer to consider the permanent load only, therefore the following rule.

SECTIONAL AREAS OF PISTON RODS IN SQUARE

e piston's diameter

by the initial pressure, that result must have added to it the permanent load in pounds, then

rces;

next, those forces combined, divided by the working strain of the material used, equals the area of the rod.

It is usual to use a factor of safety at the end of this calculation, and the number of that factor ranges according to the working strain of the metal—as, for example, should one-tenth of the breaking strain be the working strain, then the factor of safety at the end of calculation

,

therefore, can be arranged as follows:—Area of

$$\left[A \times P\right] \left[+ L = F\right] \frac{F}{B} \times S = \text{area of rod.}$$

The value of S ranges from 8 to 12. We prefer 10 as a constant.

We here give a table of the strains to which the moving details are subjected.

TABLE OF STRAINS THE DETAILS OF COMPOUND-ENGINES ARE SUBJECTED TO.

NAME OF DETAIL.	NATURE OF STRAIN.
Piston Rods.	Compression and Tensile.
Connecting Rods.	Compression and Tensile.
Securing Bolts.	Compression and Tensile.
Guide Block Pin.	Shearing.
Slide Valve Rods.	Compression and Tensile.
Crank Shaft.	Torsion and Shearing.

It will be therefore observed that the preceding rule for the area of rods applies also to the securing bolts—in fact, it must be so if the strains are alike, and the only difference in the guide block pin is the nature of the strain.

Now in the case of the crank shaft the nature of the strain is entirely at right angles to the compression and tensile strains.

It has long been proved that the strengths of round shafts are in proportion to the cube of the diameter, and it is for that reason that the sectional area is not mentioned in the formula.

DIAMETER OF REVOLVING OR CRANK SHAFTS.—We here, also, must consider the permanent load again, but with this difference, we leave it out in this formula because the rods and bolts have taken up all the shock of momentum, and it is the residue of the power that drives the crank shaft; we, therefore, understand that that power is put in the crank pin, and as the crank is a lever we must consider its length in our calculation. Then let area of piston A, initial pressure P, force F, throw or length of crank C, breaking strain B, cube of diameter D, safety factor S:

$$\left[A \times P \times C = F\right] \quad \left[\frac{F}{B} = D\right] \quad \left[\sqrt[3]{D \times S}\right]$$

= diameter of the shaft.

The diameter of the crank pin should be made equal to the diameter of the shaft, but more for the purposes of manufacture and strength than proportion. The length of the crank pin should never be less than the diameter of the pin. The sectional shape of each crank should be very nearly square; if there be any difference from that shape, each crank in width may be about three-fourths of the shaft's diameter.

The length of bearing for the shaft should be twice its diameter, but as circumstances of arrangement very often affect this proportion, the intermediate bearings are often shorter.

Surface Condenser.—The main point to be considered here is the superfiicial area of the tubes in connection with the temperature of the exhausting steam and the temperature of the circulating water. We have given this subject a deal of attention, and are thereby enabled to condense this matter into a practical table, as shown.

TABLE OF THE RATIOS OF TUBE SURFACES FOR SURFACE CONDENSERS.

Temperature of the Circulating Water in degrees Fahrenheit.	Temperature of the Exhaust Steam entering the Condenser in degrees Fahrenheit	Ratios of Tube Surface in square feet, per indicated horse power.	No. of Cubic inches of Circulating Water per stroke of the Pump, per square ft. of tube surface.
60	216·3	·25	5·96
60	219·5	·462	5·60
60	222·5	·662	5·27
60	225·4	·856	4·95
60	228·0	1·030	4.66
60	230·6	1·204	4·37
60	233·1	1·370	4·10
60	235·5	1·520	3·84
60	237·9	1·690	3·52
60	240·2	1·844	3·30
60	242·3	1·982	3·07
60	244·4	2·122	2·84
60	246·4	2·256	2·61
60	248·4	2·390	2·39
60	250·4	2·516	2·18
60	252·2	2·642	1·97
60	254·1	2·770	1·76
60	255·9	2·890	1·56
60	257·6	3·002	1·37
60	259·3	3·116	1·18

AREA OF CONDENSER TUBE SURFACE IN SQUARE FEET.—Multiply the ratio of the tube surface A by the indicated horse-power collectively I equal area of tube surface S:

$$\left[A \times I = S\right]$$

CUBICAL CAPACITY OF AIR PUMP IN INCHES.—Divide the capacity of the low pressure cylinder in inches by 6 to 7, those constants having been proved to be sufficient in practice for single acting pumps.

Of course when double action is used the capacity of the pump can be reduced by one-half.

CUBICAL CAPACITY OF CIRCULATING PUMP IN INCHES.—Multiply the area of the tube surface in square feet T by number of cubic inches of water, W, according to the temperature of the steam when entering the condenser. Then the cubical contents in inches C, divided by the length of the stroke of the piston in inches L equals the area of the piston in square inches P.

$$\left[T \times W = C\right] \quad \left[\frac{C}{L} = P\right]$$

CHAPTER XII.

SYSTEMATIC STEAM FORMULÆ.

FORMULÆ TO OBTAIN THE PROPORTIONS OF A COMPOUND ENGINE OF 1,510 INDICATED HORSE POWER, HAVING TWO CYLINDERS (HIGH AND LOW PRESSURE) SIDE BY SIDE, HORIZONTAL DIRECT-ACTING SINGLE PISTON RODS.—Proportions and constants thus:—

Indicated horse power collectively 1,510. Initial pressure of steam 80 lbs. Stroke of piston 4 feet 6 inches. Length of cut off 18 inches. Unit of heat constant 1·678. Steam constant 2·863. Temperature of the steam 324·1. Weight of 1 cubic foot of steam ·2198·

AREA OF HIGH PRESSURE CYLINDER:—

$$\left[\frac{1510}{1{\cdot}678} = 899{\cdot}88\right] \left[\frac{899{\cdot}88}{324{\cdot}1} = 2{\cdot}77655\right]$$

$$\left[\frac{2{\cdot}77655}{{\cdot}2198} = 12{\cdot}632\right] \left[12.632 \times 1728 = 21828{\cdot}096\right] \left[\frac{21828{\cdot}096}{18} = 1212{\cdot}672 = \right.$$

Area of high pressure cylinder = $39\frac{5}{16}$ bare diameter.

AREA OF LOW PRESSURE CYLINDER :—1212·672 × 4 = 4850·688 = 78$\frac{8}{16}$ full diameter for the low pressure cylinder.

MEAN PRESSURE IN HIGH PRESSURE CYLINDER :—

$\left[\frac{54}{18} = 3\right]$ $\left[\frac{80}{3} = 26{\cdot}666\right]$ [hyp. log. of 3 = 1·098 + 1 = 2·098 × 26·666 = 55·945 = theoretical mean pressure in high pressure cylinder.

MEAN PRESSURE IN LOW PRESSURE CYLINDER :—

$\left[\frac{80}{3} = 26{\cdot}666\right]$ $\left[\frac{54}{4} = 13{\cdot}5\right]$ $\left[\frac{54}{13{\cdot}5} = 4\right]$

$\left[\frac{26{\cdot}666}{4} = 6{\cdot}666\right]$ [hyp. log. of 4 = 1·386 + 1 = 2·386 × 6·666 = 15·905 = theoretical steam mean pressure in low pressure cylinder. Then if a vacuum of say 12 lbs. is obtained, the mean pressure will be 15·905 + 12 = 27·905 vacuum and steam mean pressure collectively.

ACTUAL MEAN PRESSURE IN BOTH CYLINDERS IN CONNECTION WITH THE STEAM CONSTANT :—

$\left[55{\cdot}945 + 27{\cdot}905 = 83{\cdot}850\right]$ $\frac{83{\cdot}850}{2{\cdot}863} =$ 29·28 as the collective working mean pressure.

Then—

$$\frac{55{\cdot}94}{1{\cdot}4315} = 39{\cdot}08$$ as the working mean pressure in the high pressure cylinder.

And—

$$\frac{27{\cdot}905}{1{\cdot}4315} = 19{\cdot}49$$ as the working mean pressure in the low pressure cylinder.

Next—

$$39{\cdot}08 + 19{\cdot}49 = \frac{58{\cdot}57}{2} = 28{\cdot}57$$ as the mean sum of the two pressures combined, which act as a check on the half constant 1·4315 used as a divisor to produce the sums of the working mean pressures separately.

SPEED OF THE PISTON FROM THE UNITS OF HEAT:—

[1212·672 × 39·08 = 47391·22176] [4850·688 × 19·49 = 94539·90912] [47391·22176 + 94539·90912 = 141931·13088] = the compound steam power, and

$$\frac{141931{\cdot}13088}{899{\cdot}88} = 157{\cdot}72$$ as the unit power constant, which would be required in the absence of the compound steam power as a multiplier for the units of heat to produce the power, but as we have obtained that from the area of the cylinders and

the mean pressures, the proceeding formula can be thus:—

$$1510 \times 33000 = \frac{49830000 \cdot 000}{141931 \cdot 13088} = 351 \cdot 08$$ as the actual speed of the piston obtainable.

INDICATED HORSE POWER OF HIGH PRESSURE CYLINDER:—

$$1212 \cdot 672 \times 39 \cdot 08 = 47391 \cdot 22176$$

$$47391 \cdot 22176 \times 351 \cdot 08 = 16638110 \cdot 135508$$

$$\frac{16638110 \cdot 135508}{33000} = 504 \cdot 18515 = \text{I.H.P.}$$

INDICATED HORSE POWER OF LOW PRESSURE CYLINDER:—

$$4850 \cdot 688 \times 19 \cdot 49 = 94539 \cdot 90912$$

$$94539 \cdot 90912 \times 351 \cdot 08 = 33191071 \cdot 2938496$$

$$\frac{33191071 \cdot 2938496}{33000} = 1005 \cdot 79003 = \text{I.H.P.}$$

Then—

$$1005 \cdot 79003 + 504 \cdot 18515 = 1509 \cdot 97518 = \text{I.H.P.}$$ collectively.

This result proves also that with a given pressure and a certain expansion the speed can be regulated while the indicated horse power may be a constant, as can be seen by the table on page 153 as a comparison from practice.

We may here remark that a very high rate of piston is not conducive to economy with compound engines, and therefore the calculation from the formulæ given, produce the requisite speed in observance of that fact.

We may remark also that the formulæ for the units of heat are dependent on its constant, which in the present example is moderate, for the purpose of ensuring a full power from a high grade of expansion and a moderate speed. In accordance with this also is the steam constant full, from which fact the mean pressures are less, showing thereby a loss of heat as shown in practice in connection with the units allowed.

Now, should this example have cylinders enclosed entirely with high temperature unignited gas and the steam used in the cylinders be properly made first and equally well used after, then the unit of heat constant will be increased from 1·678 to 2·00, and the steam constant decreased from 2·863 to 2·00 at the most or even 1·86. The initial pressure being the same, and the grade of expansion similar, as also the indicated horse power, with a lesser consumption of steam to equal that power, and consequently require smaller boilers and less fuel.

Let us recapitulate: the "unit of heat constant" is used as a divisor into the indicated horse power required, therefore the larger that constant the smaller sum is the units of heat, and from that fact are the areas of both or more cylinders of the engine reduced also.

But in the case of the use of the "steam constant," as it is a divisor to show the loss of the heat, the larger that sum as the divisor is, the greater the loss will be also.

Now this conclusion may appear paradoxical, but if we refer to the calculations in chapter 10 we shall see the force of our conclusion, for there it is shown that the theoretical mean pressure that ought to have driven the engine is used as a divisor into the working mean pressure, thereby producing the steam constant. Now if this steam constant is used as a divisor into the working mean pressure it produces the theoretical mean pressure, and if that sum is deducted from the sum of the working mean pressure, it will show the loss of pressure and likewise the loss of heat.

It has been shown also, that if the compound steam power is divided by the result of the multiplication of the two constants, viz., that are the indicated power required and the 33000, then the speed of the piston will be in proportion to the units of heat per stroke.

It may be mentioned that the nearer the sums of the constants for heat and steam are, the better the economy must result in practice. This conclusion may seem singular but it is a fact.

Scientific Table of Constants to obtain the correct proportions of High and Low Pressure Cylinders for Compound Engines of Modern Practice.

Collective Indicative Horse Power.	Initial Steam Unit of Heat Constants.	Loss of heat. Steam Constants.	Remarks.
100 to 150	1·000 to 1·010	3·455 to 3·386	The object of this Table is to give a practical conclusion of the proportion of the cylinders for High and Low pressure steam combined. Initial Steam, ranging from 120lbs. to 80lbs. on the square inch, the latter being used for the larger engines. The cut-off is one-third, and the constants are in conjunction with these facts. The exhaust pressure of the steam from the low pressure cylinder ranges from 10lbs. to 7lbs. The speed of the piston is from 350 feet to 500 feet. Steam jacketing is not considered in this case, but lagging is fully allowed for. The formula for the adaption of the constants is practically shown in Chaps. 9, 10, and 12, also some of the results can be seen from the tables therein and in Chap. 11.
200 ,, 255	1·067 ,, 1·121	3·317 ,, 3·263	
310 ,, 370	1·172 ,, 1·220	3·209 ,, 3·169	
430 ,, 495	1·265 ,, 1·307	3·129 ,, 3·091	
560 ,, 630	1·346 ,, 1·383	3·053 ,, 3·025	
700 ,, 775	1·418 ,, 1·451	2·997 ,, 2·977	
850 ,, 930	1·483 ,, 1·514	2·957 ,, 3·941	
1010 ,, 1100	1·544 ,, 1·573	2·925 ,, 2·912	
1190 ,, 1290	1·601 ,, 1·620	2·899 ,, 2·888	
1390 ,, 1510	1·654 ,, 1·678	2·876 ,, 2·863	
1630 ,, 1780	1·703 ,, 1·729	2·849 ,, 2·834	
1930 ,, 2110	1·756 ,, 1·784	2·818 ,, 2·801	
2290 ,, 2510	1·813 ,, 1·843	2·783 ,, 2·764	
2730 ,, 3000	1·875 ,, 1·909	2·744 ,, 2·723	
3270 ,, 3600	1·946 ,, 1·986	2·701 ,, 2·678	
3930 ,, 4330	2·030 ,, 2·078	2·654 ,, 2·629	
5230 ,, 5730	2·130 ,, 2·186	2·603 ,, 2·576	
6230 ,, 6880	2·240 ,, 2·304	2·548 ,, 2·519	
7530 ,, 8355	2·372 ,, 2·444	2·489 ,, 2·458	
9180 ,,10205	2·520 ,, 2·601	2·426 ,, 2·393	

STATISTICAL TABLE OF THE WORKING RESULTS OF MODERN COMPOUND ENGINES.

Name of Ship.	Indicated Horse Power.	Collective Area of Cylinders.	Initial Steam Pressure in lbs. per square inch.	Working Mean Pressure in High Pressure Cylinder.	Working Mean Pressure in Low Pressure Cylinder.	Working Mean Pressure Collectively.	Theoretical Mean Pressure Collectively.	Steam Constant or Working Mean Pressure divided by Theoretical Mean Pressure.	Indicated Horse Power divided by Units of Heat per stroke.	Maker of Engines.
"Garonne" ...	2650	11322·326	44·3	22·10	13·80	35·90	15·82	2·26	1·64	M[illegible] N[illegible] & Sons.
"Amérique"	1617	5738·132	54·3	39·00	11·85	50·85	18·08	2·81	1·79	M[illegible] Maudslay and Sons.
"Arndt"	1550	7851·643	64·3	39 00	10·45	49·45	15·51	3·18	3·26	M[illegible] Oswald & Co
"Mongolia"	1258	9047·807	42·3	22 12	7 23	29 35	10·19	2·88	1·12	Messrs. Day, Summers & Co.
"Tir "	1234	4649·567	62·3	44·00	10·35	54·35	17·66	3·08	1·71	M[illegible] Maudslay and Sons.
"Nankin" ...	1221	4982·577	65·3	4[illegible]0	10·10	52·70	17·[illegible]3	3 07	1·33	" "
"Dhoolia" ...	1097·6	6871·445	54 3	38·00	10·00	48·00	14·48	3 31	1·15	M[illegible] Oda[illegible] 1 & Co.
"Olbers"	1062	4081·723	72·3	62·20	10·55	72·75	18·91	3·84	1·66	Messrs. R. Stephenson & Co.
"Wallace" ...	1030	6871·464	43·3	31·12	8·00	39·12	13·58	2·88	1·21	Messrs. Oswald & Co
"Danube" ...	862·3	5089·380	50·3	34·38	6 95	41·33	12 42	3·32	1·22	M[illegible] Day, S[illegible] & Co.
"Jos Baro"	752	4383·317	55·3	37·20	5·90	43·10	12 75	3·39	1·12	Messrs. Oswald & Co
"Patroclus"	732	3549·271	65·3	50·35	10 35	60·70	17 45	3·47	1·23	Messrs. R. Stephenson & Co.
"Aristocrat"	645	2261·951	70·3	61·75	11·9	73·65	21·78	3 38	1·47	" "
"Per Jex) "	619	3123·535	58·3	40·80	12·15	52·95	18·16	2·91	2 07	Messrs. Maudslay and Sons.
"Wallace" *	575	6871·430	60·3	26·70	5·40	32·10	9 86	3·25	1·92	Messrs. Oda[illegible] 1 & Co.
"Savernake"	332·6	1800·911	54 3	37·70	9·15	46·85	15·62	3·00	1·88	" "
"Normanton"	322·25	2300·436	40·3	35·27	7·38	42·65	13·34	3·19	1·30	" "
"LadyJosyan"	294	2654·651	42·3	26·30	9·30	35·60	12·68	2 80	1·34	Messrs. Day, Summers & Co.

*half power.

M

CHAPTER XIII.

BOILER FORMULÆ.

Rule for the collapsing pressure in lbs. per square inch to crush in flue tubes $= \dfrac{T^2 \times 67166}{D \times L}$

when T = thickness in inches,
,, D = diameter in feet,
,, L = length in feet.

Rule for the bursting pressure in lbs. per square inch of cylindrical boilers along the sides =

$$B = \frac{T \times t}{R}$$

when T = tensile breaking strain of the material construction, such as riveted joints,
,, t = thickness of the plate in inches,
,, R = radius of boiler's diameter in inches,
,, B = bursting pressure.

Of course the real strength of any boiler is its weakest part, and the consideration most requisite is to strengthen that part so as to equalize the strain throughout.

TABLE TO FIND THE WORKING PRESSURE FOR CYLINDRICAL BOILERS WITH A GIVEN THICKNESS OF SHELL PLATING.

Thickness of plate.	Single riveting.	Double riveting.	Thickness of plate.	Single riveting.	Double riveting.
$\frac{1}{32}$	274	346	$\frac{17}{32}$	4,715	5,894
$\frac{1}{16}$	554	693	$\frac{9}{16}$	4,992	6,241
$\frac{3}{32}$	831	1,039	$\frac{19}{32}$	5,269	6,587
$\frac{1}{8}$	1,109	1,387	$\frac{5}{8}$	5,547	6,937
$\frac{5}{32}$	1,306	1,733	$\frac{21}{32}$	5,824	7,281
$\frac{3}{16}$	1,664	2,080	$\frac{11}{16}$	6,102	7,620
$\frac{7}{32}$	1,941	2,426	$\frac{23}{32}$	6,379	7,974
$\frac{1}{4}$	2,219	2,774	$\frac{3}{4}$	6,657	8,322
$\frac{9}{32}$	2,496	3,120	$\frac{25}{32}$	6,934	8,663
$\frac{5}{16}$	2,773	3,467	$\frac{13}{16}$	7,211	9,015
$\frac{11}{32}$	3,050	3,813	$\frac{27}{32}$	7,488	9,361
$\frac{3}{8}$	3,328	4,161	$\frac{7}{8}$	7,766	9,709
$\frac{13}{32}$	3,605	4,507	$\frac{29}{32}$	8,043	10,053
$\frac{7}{16}$	3,882	4,854	$\frac{15}{16}$	8,321	10,402
$\frac{15}{32}$	4,159	5,200	$\frac{31}{32}$	8,598	10,748
$\frac{1}{2}$	4,438	5,548	1	8,876	11,096

Divide the tabular No. opposite the thickness of shell plating and under the heading of the respective class of riveting by the extreme diameter of the boiler in inches; the quotient will be the pressure in pounds per square inch at which the boiler may be worked while in good order.

	Lbs.
Pressure per square inch on solid stays to be not more than	6000
Pressure per square inch on screw stays, taking the diameter over the threads .	5000

SQUARE AND CYLINDRICAL SHELLS.

Total heating surface of the tubes = I.H.P., × 2 to 2·5 for boilers 1000 to 500 I.H.P.; and 3 to 4 for boilers 450 to 100 I.H.P.

Diameter of tubes externally = 2 to 3 inches.

Length of tubes = 5 to 7 feet.

$$\text{Number of tubes} = \frac{\text{total surface}}{\text{surface of one tube}}.$$

Rake or inclination of tubes = $\frac{1}{8}$ to $\frac{3}{4}$ of an inch per foot.

Water space = 4 to 6 inches.

Position of stays at right angles above fire boxes = 14 to 16 inches

Position of stays at sides and bottom of fire boxes = 12 to 14 inches.

Rule for the area in square inches of gussets and stays for the flat ends of cylindrical boilers when the thickness of the plates = thickness of side plates × ·6, then $T = \frac{A \times P \times F}{B}$, and

A = area of end of boiler in square inches,

P = pressure of steam in lbs. per square inch,

F = factor of safety,
B = breaking tensile strain in lbs. per square inch.
T = total area of stays and gussets in square inches.

Number of tubes to one fire box should never exceed 125.

Width of fire box at tube = pitch of tubes × number of tubes transversely.

Fire bar or grate surface = I.H.P. × ·09 to ·18, for boilers from 1000 to 100 I.H.P.

Length of fire bar grate surface = 7 feet as a maximum, 5 to 6 feet being generally adopted.

Width of fire box at grate =

$$\frac{\text{surface of grate}}{\text{length of grate surface}}$$

Radius for top and bottom curves of fire box = width of fire box.

$$\text{Radii of small curve} = \frac{\text{width of fire box}}{4 \text{ to } 5}$$

Width of fire door opening 18 inches as a minimum.

Area of fire box at grate = grate surface × ·5·

$$\text{Area of space above bridge} = \frac{\text{area of surface grate}}{4}$$

Height of water line above fire box at tube end = 6 to 8 inches.

Width of fire box at back end = 18 inches; this

will allow room for closing or riveting the end of the tubes when renewed.

Width of smoke box at bottom = 14 inches as a minimum.

Area of opening in uptake = total area of tubes as a minimum; total area of tubes × 1·25 as a maximum.

$$\text{Area of chimney} = \frac{\text{total area of grate surface}}{8 \text{ to } 11}$$

In war ships the following should be observed:—

Top of boiler should be one foot below water line as a minimum; funnel to be telescopic, raised and lowered by two chains on a barrel, keyed on a shaft, to which motion is given by a worm and wheel on each side of the funnel.

Diameter of shaft = 2 to 3½ inches.

Diameter of wheel = 18 to 24 inches.

Pitch of teeth = 1½ to 2 inches.

$$\text{Diameter of worm} = \frac{\text{diameter of wheel}}{3 \text{ to } 5}$$

Radius of handle = 14 to 16 inches.

In order to reduce the temperature between deck and the stokehole, the funnel is surrounded by two or three casings, 4 to 6 inches of spaec between each, commencing on the main or weather deck, and terminating on the orlop or lower deck;

by this means a continuous current of air passes through. The stokehole is further ventilated, and draught increased in some cases by tubes, the tops of which are termed cowls, from being enclosed semicircularly, having the opening at the side, the top being rotative, and its position subservient to the wind.

MARINE SAFETY VALVES.—These valves are mostly weighted directly, and the following are the rules:—

Area of valve in square inches =

$$\frac{\text{total area of grate's surface in feet}}{3}$$

Diameter of valve spindle = $\frac{\text{diameter of valve}}{4}$

Diameter of weight = diameter of valve × 2.

Pressure in lbs. against the valve = pressure per square inch × area of the valve.

Cubical contents of weight, including weight of valve and spindle =

$$\frac{\text{pressure in lbs. against the valve}}{\cdot 4103 \text{ if lead and } \cdot 2361 \text{ if cast iron}}$$

$$\text{Length of weight} = \frac{\text{cubic contents of weight}}{\text{area of weight}}$$

Thickness of casing = ½ to ¾ of an inch.

Depth of guide ribs of valves = diameter of valve × ·5·

Diameter of lifting lever weight shaft = diameter of valve spindle.

Length of lifting lever = diameter of weight — ½ inch for clearance between weight.

$$\text{Lift of valves} = \frac{\text{diameter of valve}}{4}$$

Spring safety valves may be used if preferred for main valves, but they must be designed with care, or the valve and lever gear will soon get dangerous from their non-action.

In every case a small lock-up dead weight safety valve and whistle should be fitted to the front, in sight, near the top of each boiler.

FIRE BARS.—Length should never exceed 3 feet 6 inches.

Inclination for marine boilers = 2 inches per foot.

Inclination for land boilers = 1 inch per foot.

Depth of bar at the centre = 1½ to 1⅜ of an inch per foot of length.

Depth of bar at ends = ¾ of an inch per foot of length.

Width of bar at ends = ¾ to 1 inch.

Taper of sides of bar = ⅛ of an inch per inch.

Clearance for ashes = ¼ to ⅜ of an inch.

Depth of centre bearing bar = depth of fire bar centre.

Width of centre bearing bar = depth of fire bar at end × 2.

Width of end bearing bar = depth of fire bar at end.

MARINE COAL BUNKERS, ETC.—Thickness of plates,

Top plates,

Bottom plates, $\frac{3}{16}$ of an inch.

i of curves, 6 to 12 inches.

Corner angle iron, $1\frac{1}{2} \times 1\frac{1}{2} \times \frac{1}{4}$.

Stay angle iron, $2 \times 2 \times \frac{5}{16}$.

Stays, 3 feet pitch.

Temperature tubes, number = 1 per 30 tons of coals, in bunkers containing above 200 tons.

Number of cubic feet per ton of coals = 46.

Space between boilers, or width of stokehole = 9 to 10 feet.

Minimum space allowed for passing behind cylinders or thrust block in screw alley = 12 inches; maximum space 18 inches.

TABLE OF THE STRENGTH OF MATERIALS USED IN BOILER-MAKING.

NAME OF MATERIAL.	TENSION Breaking Strain per square inch in lbs	TENSION. Yielding Strain per square inch in lbs.	COMPRESSION Breaking Strain per square inch in lbs. 30 diameters	COMPRESSION Yielding Strain per square inch in lbs. 30 diameters	TORSION. Breaking Strain per square inch in lbs.	TORSION	SHEARING.	SHEARING. Yielding Strain per square inch.	Practical Working Strains one-tenth of breaking strains. Lbs. on the square inch.			
									Tensile.	Compression.	Torsion.	Shearing.
Steel bars	89,600	67.200	49,280	47,040	25,497			51,968	8,960	4,928	2,549	6,496
Wrought-iron bars...	64,960	51,968	37,000	36,000	17,000			40,800	6,496	3,700	1,700	5,000
Wrought-iron plates	50,000	40.000	36.000	34,000	...			...	5,000	3,600	...	...
Cast iron	17,000	13.000	90.000	85,000	7,000			19,000	1,700	9,000	700	2,000
Gun metal............	35,000	28,000	12,000	11,000	9,000			20,000	3,500	1,200	900	2,500
Copper sheets	28.000	24,000	15,000	12,200	...			...	2,800	1,500	...	...
Copper bars	33,600	26,680	18,000	14,400	11,200			19,000	3,360	1,800	1,120	2,100
Phosphor Bronze ...	100980	...	...	...	...			...	10,098	...	...	...

TABLE FOR CALCULATING THE WEIGHT IN LBS. PER SQUARE FT. OF DIFFERENT MATERIALS IN PLATES.

Thickness in inches.	Wrought Iron.	General Steel.	Wrought Copper.	Zinc.	Lead.	Brass.
$\frac{1}{16}$	2·5	2·59	2·903	2·301	3·701	2·705

When the plate exceeds $\frac{1}{16}$th of an inch in thickness, multiply the proportional excess by the weight of the normal thickness in lbs. = the total weight in lbs.: as, for example, suppose a wrought iron plate = $\frac{3}{4}$ inch in thickness, then as $\frac{3}{4}$ inch = $\frac{12}{16}$ of an inch, 12 × 2·5 = 30 lbs. per quare foot, which will give the actual result as if from a table, without the liability of confusion.

TABLE OF THE WEIGHTS OF ANGLE IRON OF EQUAL SIDES.

Width of each side in inches.	Thickness in inches at root.	Thickness in inches near edge	Weight in lbs. per lineal foot.
1½	5/16 bare	¼ bare	2·653
1¾	5/16	¼	3·251
2	5/16 full	¼ full	3·874
2¼	7/16	5/16 full	5·011
2½	½	⅜	6·512
2¾	9/16	9/16	8·251
3	⅝	½	10·381
3½	⅝	½ full	12·101
4	11/16	3/16	14·561

Table of the Weight of a Lineal Foot of Round and Square Bar Iron in lbs.

Diam. or side.	Square Bars.	Round Bars.	Diam. or side.	Square Bars.	Round Bars.	Diam. or side.	Square Bars.	Round Bars.
1/4	·209	·164	1 1/4	5·25	4·09	3	30·07	23·60
5/16	·326	·256	1 3/8	6·35	4·96	3 1/4	35·28	27·70
3/8	·470	·369	1 1/2	7·51	5·90	3 1/2	40·91	32·13
7/16	·640	·502	1 5/8	8·82	6·92	3 3/4	46·97	36·89
1/2	·835	·656	1 3/4	10·29	8·03	4	53·44	41·97
9/16	1·075	·831	1 7/8	11·74	9·22	4 1/4	60·32	47·38
5/8	1·305	1·025	2	13·36	10·49	4 1/2	67·63	53·12
11/16	1·579	1·241	2 1/8	15·08	11·84	4 3/4	75·35	59·18
3/4	1·879	1·476	2 1/4	16·91	13·27	5	82·51	65·58
13/16	2·205	1·732	2 3/8	18·84	14·79	5 1/4	93·46	72·30
7/8	2·556	2·011	2 1/2	20·87	16·39	5 1/2	101·03	79·35
15/16	2·936	2·306	2 5/8	23·11	18·07	5 3/4	110·43	86·73
1	2·340	2·620	2 3/4	25·26	19·84	6	120·24	94·43
1 1/8	4·220	3·320	2 7/8	27·61	21·68	7	152·00	121·4

To convert into weight of other metals, multiply tabular No. for cast iron by ·93, for steel × 1·02, for copper × 1·15, for brass × 1·09, for lead × 1·47, for zinc × ·92.

CHAPTER XIV.

GENERAL DATA AND TABLES.

Data of Specific Gravities.

	Weight of a Cubic Inch in Lbs.
Copper, Cast - - -	·3178
Iron, Cast - - -	·2631
Iron, Wrought - -	·2756
Lead - - - -	·4103
Steel - - - -	·2827
Gun Metal - - -	·3177

Data of Gravity of Water.

1 cubic foot - -	=	6·25 imperial gallons.
11·2 imperial gallons	=	1 cwt.
224 „	=	1 ton.
1 cubic foot of sea water	=	64·2 lbs.
34·9 „ „	=	1 ton.
277·274 cubic inches	=	1 imperial gallon.
1 gallon of fresh water	=	10 lbs.
1 gallon of sea water	=	10·25 lbs.

DATA OF HEAT-CONDUCTING POWER OF METALS.

Copper - - - -	1,000
Brass - - - -	468
Wrought Iron - - -	336
Cast Iron - - -	311
Lead - - - -	161
Brick - - - -	10

DATA OF THE TEMPERATURES IN DEGREES FAHR. WHEN CERTAIN MATERIALS MELT.

Wrought iron - - -	3,800
Cast Iron - - -	3,350
Copper - - - -	2,600
Brass - - - -	2,000
Zinc - - - -	700
Lead - - - -	599

PROPORTIONS OF A CIRCLE.

Diameter of a circle × 3·1416 = the circumference.
Circumference „ × ·31831 = the diameter.
Diameter „ × ·8862 = the side of an equal square.

Diameter „ × ·7071 = the side inscribed square.

Side of a square × 1·128 = the diameter of an equal square.

Square of diameter × ·7854 = the area of the circle.

Square root of area × 1·12837 = the diameter of equal circle.

Surfaces and Solids.

Square of the diameter of a sphere × 3·1416 = convex surface.

Cube of the diameter of a sphere × ·5236 = the solidity.

Diameter of a sphere × ·806 = dimensions of equal cube.

Diameter of a sphere × ·6667 = length of equal cylinder.

Square inches × ·00695 = square feet.

Cubic inches × ·00058 = cubic feet.

Cubic feet × ·03704 = cubic yards.

Circular inches × ·00456 = square feet.

Cylindrical inches × ·0004546 = cubic yards.

Cylindrical feet × ·02909 = cubic yards.

Lineal feet × ·00019 = English miles.
Lineal yards × ·000568 = English miles.
Square yards × ·0002067 = English acres.

Measures and Weights.

1728 × 1 inch = 1 cubic foot.
183·346 circular inches = 1 square foot.
22·00 cylindrical inches =1 square foot.
Cubic feet × 6.232 = imperial gallons.
Cubic inches × ·003607 =imperial gallons.
French metres × 3·281 = English feet.
,, litres × ·2202 = imperial gallons.
,, grammes × ·002205 = avoirdupois lbs.
,, kilogrammes × 2·205 =avoirdupois lbs.
Avoirdupois lbs. × ·009 = cwts.
Avoirdupois lbs. × ·00045 = tons.

Algebraic Signs as applied in Mechanical Calculations.—

= Sign of equality, and signifies equal to, as 2 added to 5 = 7.

\+ Sign of addition, and signifies plus or more, as 4 + 2 = 6.

\- Sign of subtraction, and signifies minus or less, as 7 − 5 = 2.

$\times$ Sign of multiplication, and signifies multiplied by, as $7 \times 6 = 42$.

$\div$ Sign of division, and signifies divided by, as $20 \div 5 = 4$.

$\sqrt{}$ Sign of square root
$\sqrt[3]{}$ Sign of cube root
} evolution, or the extraction of roots, thus $\sqrt[2]{81} = 9$ $\sqrt[3]{729} = 9$.

Fractions of a Foot in Inches.	Decimal Value in feet.	Area in feet	Circumference in feet.
11	·9166	·6598	2·879
10	·8333	·54537	2·617
9	·75	·44178	2·356
8	·6666	·33799	2·094
7	·5833	·26722	1·832
6	·5	·19635	1·57
5	·4166	·1363	1·308
4	·3333	·08724	1·047
3	·25	·04908	·7854
2	·1666	·02179	·5233
1	·0833	·00544	·2616
$\frac{7}{8}$	·07291	·00417	·22907
$\frac{3}{4}$	·0625	·00306	·19635
$\frac{5}{8}$	·05208	·0028	·16362
$\frac{1}{2}$	·04166	·00136	·130899
$\frac{3}{8}$	·03125	·00076	·098174
$\frac{1}{4}$	·02083	·00035	·06545
$\frac{1}{8}$	·01041	·000085	·032719

Fractions of an Inch.	Decimal Value.	Fractions of an Inch.	Decimal Value.
$\frac{7}{8}$ & $\frac{3}{32}$	·96875	$\frac{3}{8}$ & $\frac{3}{32}$	·46875
$\frac{7}{8}$ & $\frac{1}{16}$	·9375	$\frac{3}{8}$ & $\frac{1}{16}$	·4375
$\frac{7}{8}$ & $\frac{1}{32}$	·90625	$\frac{3}{8}$ & $\frac{1}{32}$	·40625
$\frac{7}{8}$	·875	$\frac{3}{8}$	·375
$\frac{3}{4}$ & $\frac{3}{32}$	·84375	$\frac{1}{4}$ & $\frac{3}{32}$	·34375
$\frac{3}{4}$ & $\frac{1}{16}$	·8125	$\frac{1}{4}$ & $\frac{1}{16}$	·3125
$\frac{3}{4}$ & $\frac{1}{32}$	·78125	$\frac{1}{4}$ & $\frac{1}{32}$	·28125
$\frac{3}{4}$	·75	$\frac{1}{4}$	·25
$\frac{5}{8}$ & $\frac{3}{32}$	·71875	$\frac{1}{8}$ & $\frac{3}{32}$	·21875
$\frac{5}{8}$ & $\frac{1}{16}$	·6875	$\frac{1}{8}$ & $\frac{1}{16}$	·1875
$\frac{5}{8}$ & $\frac{1}{32}$	·65625	$\frac{1}{8}$ & $\frac{1}{32}$	·15625
$\frac{5}{8}$	·625	$\frac{1}{8}$	·125
$\frac{1}{2}$ & $\frac{3}{32}$	.59375	$\frac{3}{32}$	·09375
$\frac{1}{2}$ & $\frac{1}{16}$	.5625	$\frac{1}{16}$	·0625
$\frac{1}{2}$ & $\frac{1}{32}$	.53125	$\frac{1}{32}$	·03125
$\frac{1}{2}$	.5		

Pressure above the Atmosphere.	Sensible Temperature in Fahrenheit Degrees.	Total Heat in Degrees from Zero of Fahrenheit.	Weight of One Cubic Foot of Steam.	Relative Volume of the Steam Compared with the Water from which it was raised.
Lb.	Deg.	Deg.	Lb.	
1	216·3	1179·4	·0411	1515
2	219 6	1180·3	·0435	1431
3	222·4	1181·2	·0459	1357
4	225·3	1182·1	·0483	1290
5	228·0	1182·9	·0507	1229
6	230·6	1183·7	·0531	1174
7	233·1	1184·5	·0555	1123
8	235·5	1185·2	·0580	1075
9	237·8	1185·9	·0601	1036
10	240·1	1186·6	·0625	996
11	242·3	1187·3	·0650	958
12	244·4	1187·8	·0673	926
13	246·4	1188·4	·0696	895
14	248·4	1189·1	·0719	866
15	250·4	1189·8	·0743	838
16	252·2	1190·4	·0766	813
17	254·1	1190·9	·0789	789
18	255·9	1191·5	·0812	767
19	257·6	1192·0	·0835	746
20	259·3	1192·5	·0858	726

Pressure above the Atmosphere.	Sensible Temperature in Fahrenheit Degrees.	Total Heat in Degrees from Zero of Fahrenheit.	Weight of One Cubic Foot of Steam.	Relative Volume of the Steam Compared with the Water from which it was raised
Lb.	Deg.	Deg.	Lb.	
21	260·9	1193·0	·0881	707
22	262·6	1193·5	·0905	688
23	264·2	1194·0	·0929	671
24	265·8	1194·5	·0952	655
25	267·3	1194·9	·0974	640
26	268·7	1195·4	·0996	625
27	270·2	1195·8	·1020	611
28	271·6	1196·2	·1042	598
29	273·0	1196·6	·1065	585
30	274·4	1197·1	·1089	572
31	275·8	1197·5	·1111	561
32	277·1	1197·9	·1133	550
33	278·4	1198·3	·1156	539
34	279·7	1198·7	·1179	529
35	281·0	1199·1	·1202	518
36	282·3	1199·5	·1224	509
37	283·5	1199·9	·1246	500
38	284·7	1200·3	·1269	491
39	285·9	1200·6	·1291	482
40	287·1	1201·0	·1314	474

Pressure above the Atmosphere.	Sensible Temperature in Fahrenheit Degrees.	Total Heat in Degrees from Zero of Fahrenheit.	Weight of One Cubic Foot of Steam.	Relative Volume of the Steam Compared with the Water from which it was raised.
Lb.	Deg.	Deg.	Lb.	
41	288·2	1201·3	·1336	466
42	289·3	1201·7	·1364	458
43	290·4	1202·0	·1380	451
44	291·6	1202·4	·1403	444
45	292·7	1202·7	·1425	437
46	293·8	1203·1	·1447	430
47	294·8	1203·4	·1469	424
48	295·9	1203·7	·1493	417
49	296·9	1204·0	·1516	411
50	298·0	1204·3	·1538	405
51	299·0	1204·6	·1560	399
52	300·0	1204·9	·1583	393
53	300·9	1205·2	·1605	388
54	301·9	1205·5	·1627	383
55	302·9	1205·8	·1648	378
56	303·9	1206·1	·1670	373
57	304·8	1206·3	·1692	368
58	305·7	1206·6	·1714	363
59	306·6	1206·9	·1736	359
60	307·5	1207·2	·1759	353

Pressure above the Atmosphere.	Sensible Temperature in Fahrenheit Degrees.	Total Heat in Degrees from Zero of Fahrenheit.	Weight of One Cubic Foot of Steam.	Relative Volume of the Steam Compared with the Water from which it was raised.
Lb.	Deg.	Deg.	Lb.	
61	308·4	1207·4	·1782	349
62	309·3	1207·7	·1804	345
63	310·2	1208·0	·1826	341
64	311·1	1208·3	·1848	337
65	312·0	1208·5	·1869	333
66	312·8	1208·8	·1891	329
67	313·6	1209·1	·1913	325
68	314·5	1209·4	·1935	321
69	315·3	1209·6	·1957	318
70	316·1	1209·9	·1980	314
71	316·9	1210·1	·2002	311
72	317·8	1210 4	·2024	308
73	318·6	1210·6	·2044	305
74	319·4	1210·9	·2067	301
75	320·2	1211·1	·2089	298
76	321·0	1211·3	·2111	295
77	321·7	1211·5	·2133	292
78	322 5	1211·8	·2155	289
79	323·3	1212·0	·2176	286
80	324·1	1212·3	·2198	283

Pressure above the Atmosphere.	Sensible Temperature in Fahrenheit Degrees.	Total Heat in Degrees from Zero of Fahrenheit.	Weight of One Cubic Foot of Steam.	Relative Volume of the Steam Compared with the Water from which it was raised.
Lb.	Deg.	Deg.	Lb.	
81	324·8	1212·5	·2219	281
82	325·6	1212·8	·2241	278
83	326·3	1213·0	·2263	275
84	327·1	1213·2	·2285	272
85	327·9	1213 4	·2307	270
86	328·5	1213·6	·2329	267
87	329·1	1213·8	·2351	265
88	329·9	1214·0	·2373	262
89	330·6	1214·2	·2393	260
90	331·3	1214·4	·2414	257
91	331·9	1214·6	·2435	255
92	332·6	1214·8	·2456	253
93	333·3	1215·0	·2477	251
94	334·0	1215·3	·2499	249
95	334·6	1215·5	·2521	247
96	335·3	1215·7	·2543	245
97	336·0	1215·9	·2564	243
98	336·7	1216·1	·2586	241
99	337·4	1216·3	·2607	239
100	338·0	1216·5	·2628	237

Pressure above the Atmosphere.	Sensible Temperature in Fahrenheit Degrees.	Total Heat in Degrees from Zero of Fahrenheit.	Weight of One Cubic Foot of Steam.	Relative Volume of the Steam Compared with the Water from which it was raised.
Lb.	Deg.	Deg.	Lb.	
101	338·6	1216·7	·2649	235
102	339·3	1216·9	·2674	233
103	339·9	1217·1	·2696	231
104	340·5	1217·3	·2738	229
105	341·1	1217·4	·2759	227
106	341·8	1217·6	·2780	225
107	342·4	1217·8	·2801	224
108	343·0	1218·0	·2822	222
109	343·6	1218·2	·2845	221
110	344·2	1218·4	·2867	219
111	344·8	1218·6	·2889	217
112	345·4	1218·8	·2911	215
113	346·0	1218·9	·2933	214
114	346·6	1219·1	·2955	212
115	347·2	1219·3	·2977	211
116	347·8	1219·5	·2999	209
117	348 3	1219·6	·3020	208
118	348·9	1219·8	·3040	206
119	349·5	1220·0	·3060	205
120	350·1	1220·2	·3080	203

Pressure above the Atmosphere.	Sensible Temperature in Fahrenheit Degrees.	Total Heat in Degrees from Zero of Fahrenheit.	Weight of One Cubic Foot of Steam.	Relative Volume of the Steam Compared with the Water from which it was raised.
Lb.	Deg.	Deg.	Lb.	
121	350·6	1220·3	·3101	202
122	351·2	1220·5	·3121	200
123	351·8	1220·7	·3142	199
124	352·4	1220·9	·3162	198
125	352·9	1221·0	·3184	197
126	353·5	1221·2	·3206	195
127	354·0	1221·4	·3228	194
128	354·5	1221·6	·3250	193
129	355·0	1221·7	·3273	192
130	355·6	1221·9	·3294	190
131	356·1	1222·0	·3315	189
132	356·7	1222·2	·3336	188
133	357·2	1222·3	·3357	187
134	357·8	1222·5	·3377	186
135	358·3	1222·7	·3397	184
140	361·0	1223·5	·3500	179
145	363·4	1224·2	·3607	174
150	366·0	1224·9	·3714	169
155	368·2	1225·7	·3821	164
160	370·8	1226·4	·3928	159

Pressure above the Atmosphere.	Sensible Temperature in Fahrenheit Degrees.	Total Heat in Degrees from Zero of Fahrenheit.	Weight of One Cubic Foot of Steam.	Relative Volume of the Steam Compared with the Water from which it was raised.
Lb.	Deg.	Deg.	Lb.	
165	372·9	1227·1	·4035	155
170	375·3	1227·8	·4142	151
175	377·5	1228·5	·4250	148
180	379·7	1229·2	·4357	144
185	381·7	1229·8	·4464	141
195	386·0	1231·1	·4668	135
205	389·9	1232·3	4872	129
215	393·8	1233·5	·5072	123
225	397·5	1234·6	·5270	119
235	401·1	1235·7	·5471	114
245	404·5	1236·8	·5670	110
255	407·9	1237·8	·5871	106
265	411·2	1238·8	·6070	102
275	414·4	1239·8	·6268	99
285	417·5	1240·7	·6469	96
335	430·1	1252·3	·6643	83
385	444·9	1266·8	·6921	73
435	456·7	1277·6	·7200	66
485	467·5	1286·5	·7456	59
585	487·0	1305·7	·7681	50

Pressure above the Atmosphere.	Sensible Temperature in Fahrenheit Degrees.	Total Heat in Degrees from Zero of Fahrenheit.	Weight of One Cubic Foot of Steam.	Relative Volume of the Steam Compared with the Water from which it was raised.
Lb.	Deg.	Deg.	Lb.	
685	504·1	1321·3	·7842	43
785	519·5	1357·7	·9010	38
885	533·6	1349·5	·9231	34
985	546·5	1361·5	·9400	31
1000	600·6	1414·8	·9682	26
1500	750·8	1550·8	1·0928	19

THE uses of the steam tables have been applied in the heat calculations forming pages 72 to 107 inclusive, wherein the weight and sensible temperature are copied from the tables.

We may here explain that the reason why the sensible temperature is termed "foot degrees" in the calculations in chapter 9 is, because the total heat degrees are not considered of practical utility to obtain the units of heat in the quantity of initial steam used per stroke. But should the total heat degrees be required to be used, their sum must be divided by the foot degrees, and the constant obtained must be used as a multiplier for the steam constant C in chapter 10, from which will be obtained the sum of the total loss of heat from zero. We may add in passing that the term "foot degrees" is also obtained from the fact that a certain amount of "energy of heat" having been used to obtain the temperature according to the pressure, we were bound to notice that fact in our calculations.

The last column is the proportions of the relative volumes, and it will be seen that the higher the temperature the greater the proportionate amount of water is required to produce the relative pressure; those proportionates therefore of the volumes must be used as follows:—

$$\frac{\text{Amount of initial steam used per stroke in cubic ins.}}{\text{Relative sum of volume as per table}}$$

equals the amount of water in cubic inches required per stroke in the boiler, supposing there were no loss any way. But as there is a loss it is usual to use a multiplier of from four to six for practical purposes, from which the cubical contents of the feed pump in inches can be easily obtained.

TABLE OF CONSTANTS TO FIND THE INDICATED HORSE POWER WHEN THE SPEED OF THE PISTON AND MEAN PRESSURE OF THE STEAM ARE GIVEN.

Area of Cylinder's Diameter in inches.	Constant.	Area of Cylinder's Diameter in inches.	Constant.
7·068	·000214	22·690	·000687
7·669	·000232	23·758	·000719
8·295	·000251	24·850	·000753
8·946	·000271	25·967	·000786
9·621	·000291	27·108	·000821
10·320	·000312	28·274	·000856
11·044	·000334	29·464	·000892
11·793	·000357	30·679	·000929
12·566	·000380	31·919	·000967
13·364	·000404	33·183	·001005
14·186	·000429	34·471	·001044
15·033	·000455	35·784	·001084
15·904	·000481	37·122	·001124
16·800	·000509	38·484	·001166
17·720	·000536	39·871	·001208
18·665	·000565	41·282	·001250
19·635	·000595	42·718	·001294
20·629	·000625	44·178	·001338
21·647	·000655	45·663	·001383

Table of Constants to find the Indicated Horse Power.

Area of Cylinder's Diameter in inches.	Constant.	Area of Cylinder's Diameter in inches.	Constant.
47·173	·001429	82·516	·002500
48·707	·001476	84·540	·002526
50·265	·001523	86·590	·002623
51·848	·001571	88·664	·002686
53·456	·001619	90 762	·002750
55·088	·001669	92·885	·002814
56·745	·001719	95·033	·002879
58·426	·001770	97·205	·002945
60·132	·001822	99·402	·003012
61·862	·001874	101·623	·003079
63·617	·001927	103·869	·003140
65·396	·001981	106·139	·003216
67·200	·002036	108·434	·003285
69·029	·002091	110·753	·003356
70·882	002147	113·097	·003427
72·759	·002204	115·466	·003498
74·662	·002262	117·859	·003571
76·588	·002320	120·276	·003644
78·540	·002379	122·718	·003718
80·515	·002439	125·185	·003793

Table of Constants to find the Indicated Horse Power.

Area of Cylinder's Diameter in inches.	Constant.	Area of Cylinder's Diameter in inches.	Constant.
127·676	·003868	182·654	·005534
130·192	·003945	185·661	·005626
132·732	·004022	188·692	·005717
135·297	·004099	191·748	·005810
137·886	·004178	194·828	·005903
140·500	·004257	197·933	·005997
143·139	·004337	201·062	·006092
145·802	·004418	204·216	·006188
148·489	·004499	207·394	·006284
151·201	·004581	210·597	·006381
153·938	·004664	213·825	·006479
156·699	·004748	217·077	·006578
159·485	·004832	220·353	·006677
162·295	·004918	223·654	·006777
165·130	·005003	226·980	·006878
167·989	·005090	230·330	·006979
170·873	·005178	233·705	·007081
173·782	·005266	237·104	·007184
176·715	·005354	240·528	·007288
179·672	·005444	243·977	·007393

TABLE OF CONSTANTS TO FIND THE INDICATED HORSE POWER.

Area of Cylinder's Diameter in inches.	Constant.	Area of Cylinder's Diameter in inches.	Constant.
247·450	·007498	322·063	·009759
250·947	·007604	326·051	·009880
254·469	·007711	330·064	·010006
258·016	·007818	334·101	·010124
261·587	·007926	338·163	·010247
265·182	·008035	342·250	·010371
268·803	·008145	346·361	·010495
272·447	·008256	350·497	·010621
276·117	·008367	354·657	·010747
279·811	·008479	358·841	·010873
283·529	·008592	363·051	·011002
287·272	·008705	367·284	·011129
291·039	·008823	371·543	·011258
294·831	·008934	375·826	·011388
298·648	·009049	380·133	·011519
302·489	·009166	384·465	·011650
306·355	·009283	388·822	·011782
310·245	·009401	393·203	·011915
314·160	·009519	397·608	·012048
318·099	·009639	402·038	·012182

TABLE OF CONSTANTS TO FIND THE INDICATED HORSE POWER.

Area of Cylinder's Diameter in inches.	Constant.	Area of Cylinder's Diameter in inches.	Constant.
406·493	·012317	500·741	·015173
410·972	·012453	505·711	·015324
415·476	·012590	510·706	·015475
420·004	·012727	515·725	·015928
424·557	·012865	520·769	·015780
429·135	·013004	525·837	·015934
433·737	·013143	530·930	·016088
438·363	·013283	536·047	·016248
443·014	·013424	541·189	·016399
447·699	·013566	546·356	·016556
452·390	·013708	551·547	·016713
457·115	·013851	556·762	·016871
461·864	·013995	562·002	·017030
466·638	·014140	567·267	·017189
471·436	·014285	572·556	·017350
476·259	·014432	577·870	·017511
481·106	·014578	583·208	·017672
485·978	·014726	588·571	·017835
490·875	·014874	593·958	·017957
495·796	·015024	599·370	·018157

TABLE OF CONSTANTS TO FIND THE INDICATED HORSE POWER.

Area of Cylinder's Diameter in inches.	Constant.	Area of Cylinder's Diameter in inches.	Constant.
604·807	·018327	718·690	·021778
610·268	·018492	724·641	·021958
615·753	·018659	730·618	·022139
621·263	·018826	736·619	·022321
626·798	·018993	742·644	·022504
632·357	·019162	748·694	·022687
637·941	·019331	754·769	·022871
643·549	·019501	760·868	·023056
649·182	·019671	766·992	·023242
654·839	·019843	773·140	·023428
660·521	·020015	779·313	·023615
666·227	·020188	785·510	·023803
671·958	·020362	791·732	·023991
677·714	·020516	797·978	·024131
683·494	·020711	804·249	·024371
689·298	·020888	810·545	·024561
695·128	·021064	816·865	·024753
700·981	·021241	823·209	·024945
706·860	·021419	829·578	·025138
712·762	·021598	835·972	·025332

Table of Constants to find the Indicated Horse Power.

Area of Cylinder's Diameter in inches	Constant.	Area of Cylinder's Diameter in inches.	Constant.
842·390	·025526	975·908	·029572
848·833	·025722	982·842	·029851
855·300	·025918	989·800	·029993
861·792	·026114	996·783	·030205
868·308	·026312	1003·790	·030417
874·849	·026449	1010·822	·030630
881·415	·026709	1017·878	·030844
888·005	·026909	1024·959	·031059
894·619	·027109	1032·064	·031274
901·258	·027310	1039·194	·031490
907·922	·027512	1046·394	·031707
914·610	·027715	1053·528	·031925
921·323	·027918	1060·731	·032143
928·060	·028123	1067·959	·032362
934·822	·028262	1075·212	·032582
941·608	·028533	1082·489	·032802
948·419	·028739	1089·791	·033023
955·255	·028947	1097·117	·033245
962·115	·029143	1104·468	·033468
968·999	·029363	1111·844	·033692

Table of Constants to find the Indicated Horse Power.

Area of Cylinder's Diameter in inches.	Constant.	Area of Cylinder's Diameter in inches.	Constant.
1119·244	·033916	1272·397	·038558
1126·668	·034141	1280·312	·038797
1134·117	·034367	1288·252	·039037
1141·591	·034593	1296·216	·039279
1149·089	·034820	1304·205	·039521
1156·611	·035048	1312·219	·039764
1164·159	·035277	1320·257	·040007
1171·730	·035506	1328·320	·040252
1179·327	·035737	1336·407	·040497
1186·948	·035968	1344·518	·040742
1194·593	·036199	1352·655	·040989
1202·263	·036432	1360·815	·041236
1209·957	·036665	1369·001	·041484
1217·676	·036899	1377·211	·041734
1225·420	·037138	1385·445	·041983
1233·188	·037369	1393·704	·042233
1240·981	·037605	1401·988	·042484
1248·798	·037842	1410·296	·042736
1256·640	·038079	1418·628	·042988
1264·506	·038318	1426·985	·043241

TABLE OF CONSTANTS TO FIND THE INDICATED HORSE POWER.

Area of Cylinder's Diameter in inches.	Constant.	Area of Cylinder's Diameter in inches.	Constant.
1435·367	·043495	1608·155	·048731
1443·773	·043750	1617·042	·049001
1452·204	·044016	1625·974	·049271
1460·659	·044262	1634·920	·049542
1469·139	·044519	1643·891	·049814
1477·634	·044777	1652·886	·050037
1486·173	·045035	1661·906	·050360
1494·726	·045294	1670·950	·050634
1503·304	·045554	1680·019	·050909
1511·907	·045815	1689·103	·051185
1520·534	·046076	1698·231	·051462
1529·186	·046338	1707·373	·051738
1537·862	·046601	1716·540	·052016
1546·553	·046865	1725·732	·052294
1555·288	·047129	1734·948	·052574
1564·038	·047395	1744·189	·052854
1572·812	·047660	1753·454	·053135
1581·611	·047927	1762·734	·053416
1590·435	·048194	1772·058	·053698
1599·283	·048463	1781·397	·053981

Table of Constants to find the Indicated Horse Power.

Area of Cylinder's Diameter in inches.	Constant.	Area of Cylinder's Diameter in inches.	Constant.
1790·761	·054265	1983·184	·060096
1800·149	·054550	1993·052	·060395
1809·561	·054835	2002·966	·060556
1818·998	·055141	2012·894	·060996
1828·460	·055408	2022·846	·061348
1837·936	·055695	2032·823	·061600
1847·457	·055983	2042·825	·061903
1856·992	·056272	2052·851	·062207
1868·552	·056562	2062·902	·062512
1876·136	·056852	2072·967	·062817
1985·745	·057143	2083·077	·063123
1995·378	·057435	2093·201	·063430
1905·036	·057728	2103·350	·063737
1914·709	·058021	2113·523	·064046
1924·426	·058316	2123·721	·064355
1934·157	·058610	2133·944	·064665
1943·194	·058906	2144·191	·064975
1953·694	·059202	2154·462	·065336
1963·500	·059500	2164·758	·065598
1973·329	·059797	2175·079	·065911

TABLE OF CONSTANTS TO FIND THE INDICATED HORSE POWER.

Area of Cylinder's Diameter in inches.	Constant.	Area of Cylinder's Diameter in inches.	Constant.
2185·424	·066225	2397·482	·072651
2195·794	·066539	2408·343	·072979
2206·1886	·066854	2419·228	·073309
2216·607	·067170	2430·183	·073640
2227·050	·067486	2441·072	·073972
2237·518	·067817	2452·031	·074304
2248·011	·068121	2463 014	·074636
2258·528	·068440	2474·022	·074970
2269·069	·068760	2485·054	·075304
2279·635	·069208	2496·111	·075639
2290·226	·069400	2507·193	·075975
2300·841	·069722	2518·299	·076312
2311·481	·070044	2529·429	·076649
2322·145	·070384	2540·584	·076987
2332·834	·070691	2551·764	·077376
2343·547	·071016	2562·968	·077665
2354·285	·071342	2574·197	·078006
2365·048	·071668	2585·450	·078347
2375·835	·071994	2596·728	·078688
2386·646	·072321	2608·031	·079031

Table of Constants to find the Indicated Horse Power.

Area of Cylinder's Diameter in inches.	Constant.	Area of Cylinder's Diameter in inches.	Constant.
2619·358	·079374	2874·760	·087113
2630·709	·079718	2898·567	·087815
2642·085	·080063	2922·473	·088559
2653·486	·080408	2946·477	·089287
2664·911	·080755	2970·579	·090225
2676·360	·081100	2994·779	·090750
2687·835	·081449	3019·077	·091487
2699·333	·081798	3043·474	·092226
2710·857	·082147	3067·968	·092968
2722·405	·082497	3092·561	·093713
2733·977	·082845	3117·252	·094467
2745·574	·083199	3142·041	·095213
2757·195	·083560	3166·929	·095967
2768·841	·083904	3191·914	·096724
2780·512	·084258	3216·998	097484
2792·207	·084612	3242·178	·098247
2803·927	·084967	3267·460	·099013
2815·671	·085323	3292·838	·099783
2827·440	·085680	3318·315	·100555
2851·051	·086395	3343·887	·101300

TABLE OF CONSTANTS TO FIND THE INDICATED HORSE POWER.

Area of Cylinder's Diameter in inches.	Constant.	Area of Cylinder's Diameter in inches.	Constant.
3369·562	·102107	3903·634	·118291
3395·333	·102888	3931·368	·119132
3421·202	·103672	3959·201	·119975
3447·167	·104459	3987·130	·120823
3473·235	·105249	4015·1611	·121671
3499·398	·106042	4043·288	·122523
3525·660	·106838	4071·513	·123379
3552·018	·107636	4099·835	·124237
3578·478	·108438	4128·258	·125098
3605·035	·109243	4156·778	·125962
3631·689	·110051	4185·396	·126830
3658·440	·110866	4214·110	·127700
3685·293	·111675	4242·927	·128573
3712·242	·112492	4271·839	·12944
3739·289	·113311	4300·850	·130328
3766·432	·114134	4329·957	·131210
3793·678	·114933	4359·166	·132095
3821·020	·115788	4388·471	·132983
3848·460	·116620	4417·875	·133874
3875·996	·117454	4447·374	·134768

Table of Constants to find the Indicated Horse Power.

Area of Cylinder's Diameter in inches.	Constant.	Area of Cylinder's Diameter in inches.	Constant.
4476·976	·135665	5089·588	·154229
4506·674	·136565	5121·249	·155189
4536·470	·137468	5153·009	·156151
4566·362	·138374	5184·865	·157117
4596·357	·139283	5216·823	·158085
4626·447	·140195	5248·877	·159056
4656·636	·141110	5281·029	·160031
4686·921	·142027	5313·278	·161008
4717·308	·142948	5345·628	·161938
4747·792	·143872	5378·075	·162971
4778·373	·144795	5410·620	·163958
4809·051	·145728	5443·261	·164947
4839·831	·146661	5476·005	·165939
4870·707	·147603	5508·844	·166934
4901·681	·148535	5541·782	·167932
4932·751	·149477	5574·816	·168933
4963·924	·150421	5607·952	·169437
4995·193	·151369	5641·184	·170944
5026·560	·152320	5674·515	·171954
5058·023	·153278	5707·941	·172967

TABLE OF CONSTANTS TO FIND THE INDICATED HORSE POWER.

Area of Cylinder's Diameter in inches.	Constant.	Area of Cylinder's Diameter in inches.	Constant.
5741·470	·173983	6432·622	·194927
5775·095	·175002	6468·210	·196005
5808·818	·176024	6503·897	·197087
5842·637	·177049	6539·680	·198171
5876·559	·178078	6573·565	·199259
5910·576	·179108	6611·546	·200349
5944·692	·180142	6647·625	·201442
5978·904	·181178	6683·801	·202539
6013·218	·182218	6720·078	·203638
6047·629	·183261	6756·452	·204740
6082·137	·184307	6792·924	·205845
6116·742	·185355	6829·492	·206954
6151·449	·186407	6866·163	·208065
6186·252	·187461	6902·929	·209179
6221·153	·188519	6939·794	·210296
6256·150	·190017	6976·755	·211416
6291·250	·190687	7013·818	·212539
6326·446	·191710	7050·977	·213665
6361·740	·192779	7088·235	·214794
6397·130	·193852	7125·588	·215926

TABLE OF CONSTANTS TO FIND THE INDICATED HORSE POWER.

Area of Cylinder's Diameter in inches.	Constant.	Area of Cylinder's Diameter in inches.	Constant.
7163·044	·217061	7932·736	·240385
7200·596	·218199	7972·212	·241581
7238·246	·219340	8011·865	·242783
7275·992	·220484	8051·577	·243987
7313·841	·221631	8091·387	·245193
7351·785	·222781	8131·295	·246403
7389·828	·223933	8171·301	·247614
7427·967	·225089	8211·408	·248830
7466·208	·226248	8251·608	·250048
7504·546	·227410	8291·869	·251269
7542·981	·228574	8332·308	·252492
7581·513	·229742	8372·805	·253721
7620·147	·230913	8413·400	·254951
7658·877	·232086	8454·094	·256184
7697·705	·233263	8494·886	·257420
7736·629	·234443	8535·776	·258660
7775·656	·235625	8576·764	·259901
7814·779	·236811	8617·850	·261145
7854·000	·237999	8659·034	·262394
7893·319	·239190	8700·317	·263645

TABLE OF CONSTANTS TO FIND THE INDICATED HORSE POWER.

Area of Cylinder's Diameter in inches.	Constant.	Area of Cylinder's Diameter in inches.	Constant.
8741·698	·264899	9590·1	·290602
8783·177	·266157	9633·5	·291924
8824·754	·267416	9696·8	·293239
8866·4	·268678	9720·0	·294545
8908·2	·269945	9763·2	·295887
8950·1	·270121	9806·4	·297163
8992·0	·272485	9849·6	·298546
9034·1	·273760	9887·8	·299630
9076·3	·275038	9936·0	·301218
9118·6	·276321	9979·2	·304000
9160·9	·277602	10022·4	·303901
9203·4	·278890	10069·2	·305127
9245·9	·280179	10116·0	·306597
9288·6	·281472	10162·8	·307963
9331·3	·282767	10209·6	·309304
9374·2	·284066	10252·8	·310690
9417·1	·285367	10296·0	·312023
9460·2	·286672	10339·2	·313809
9503·3	·287979	10382·4	·314030
9546·7	·289293	10429·2	·316036

TABLE OF CONSTANTS TO FIND THE INDICATED HORSE POWER.

Area of Cylinder's Diameter in inches.	Constant.	Area of Cylinder's Diameter in inches.	Constant.
10476·0	·317497	10977·8	·332660
10522·8	·318872	11026·0	·334205
10569·6	·320252	11074·2	·335581
10616·4	·321709	11122·6	·337031
10663·2	·323019	11169·3	·338463
10710·0	·324545	11216·1	·339869
10756·8	·325797	11262·9	·341300
10800·0	·327272	11309·8	·340272
10843·2	·328588	11357·0	·344151
10886·4	·329890	11404·2	·345581
10929·6	·331390	11451·5	·347015

The use of these tables is as follows:—Required the indicated horse power of a given area of cylinder.—Multiply the constant by the velocity of the piston in feet, and that result, multiplied by the mean pressure on the piston, equals the actual indicated horse power.

TABLE OF SPEEDS.

KNOTS PER HOUR.

Sec.	3	4	5	6	7	8	9	10	11	12	13	14
	Minutes.											
0	20·000	15·000	12·000	10·000	8·571	7·500	6·667	6·000	5·455	5·000	4·615	4·286
1	19·890	14·938	11·960	9·972	8·551	7·484	6·654	5·990	5·446	4·993	4·609	4·281
2	19·780	14·876	11·921	9·945	8·531	7·469	6·642	5·980	5·438	4·986	4·604	4·276
3	19·672	14·815	11·881	9·917	8·511	7·453	6·630	5·970	5·430	4·979	4·598	4·270
4	19·565	14·754	11·842	9·890	8·491	7·438	6·618	5·960	5·422	4·972	4·592	4·265
5	19·459	14·694	11·803	9·863	8·471	7·423	6·606	5·950	5·414	4·966	4·586	4·260
6	19·355	14·634	11·765	9·836	8·451	7·407	6·593	5·941	5·405	4·959	4·580	4·255
7	19·251	14·575	11·726	9·809	8·431	7·392	6·581	5·931	5·397	4·952	4·574	4·250
8	19·149	14·516	11·688	9·783	8·411	7·377	6·569	5·921	5·389	4·945	4·569	4·245
9	19·048	14·458	11·650	9·756	8·392	7·362	6·557	5·911	5·381	4·938	4·563	4·240

TABLE OF SPEEDS.

KNOTS PER HOUR.

Sec.	3	4	5	6	7	8	9	10	11	12	13	14
	Minutes.											
10	18·947	14·400	11·613	9·730	8·372	7·347	6·545	5·902	5·373	4·932	4·557	4·235
11	18·848	14·343	11·576	9·704	8·353	7·332	6·534	5·892	5·365	4·925	4·551	4·230
12	18·750	14·286	11·538	9·677	8·333	7·317	6·522	5·882	5·357	4·918	4·545	4·225
13	18 653	14·229	11 502	9·651	8·314	7·302	6·510	5·873	5·349	4·911	4·540	4·220
14	18·557	14·173	11·465	9·626	8·295	7·287	6·498	5·863	5·341	4·905	4·534	4·215
15	18·462	14·118	11·429	9·600	8·276	7·273	6·486	5·854	5·333	4·898	4·528	4·211
16	18·367	14·032	11·392	9·574	8·257	7·258	6·475	5·844	5·325	4·891	4·522	4 206
17	18·274	14·008	11·356	9·549	8·238	7·213	6·463	5·835	5·318	4·885	4·517	4·201
18	18·182	13·953	11·321	9·524	8·219	7·229	6·452	5·825	5·310	4·878	4·511	4·196
19	18·090	13·900	11·285	9·499	8·200	7·214	6·440	5·816	5·302	4·871	4·506	4·191

TABLE OF SPEEDS.

KNOTS PER HOUR.

Sec.	3	4	5	6	7	8	9	10	11	12	13	14
	Minutes.											
20	18·000	13·846	11·250	9·474	8·182	7·200	6·429	5·806	5·294	4·865	4·500	4·186
21	17·910	13·793	11·215	9·449	8·163	7·186	6·417	5·797	5·286	4·858	4·494	4·181
22	17·822	13·740	11·180	9·424	8·145	7·171	6·406	5·788	5·279	4·852	4·489	4·176
23	17·734	13·688	11·146	9·399	8·126	7·157	6·394	5·778	5·271	4·845	4·483	4·171
24	17·647	13·636	11·111	9·375	8·108	7·143	6·383	5·769	5·263	4·839	4·478	4·167
25	17·561	13·585	11·077	9·351	8·090	7·129	6·372	5·760	5·255	4·832	4·472	4·162
26	17·476	13·534	11·043	9·326	8·072	7·115	6·360	5·751	5·248	4·826	4·466	4·157
27	17·391	13·483	11·009	9·302	8·054	7·101	6·349	5·742	5·240	4·819	4·461	4·152
28	17·308	13·433	10·976	9·278	8·036	7·087	6·338	5·732	5·233	4·813	4·455	4·147
29	17·225	13·383	10·942	9·254	8·018	7·073	6·327	5·723	5·225	4·806	4·450	4·143

TABLE OF SPEEDS.

KNOTS PER HOUR.

Sec.	3	4	5	6	7	8	9	10	11	12	13	14
	Minutes.											
30	17·143	13·333	10 909	9·231	8·000	7·059	6·316	5·714	5·217	4·800	4·444	4·138
31	17·062	13·284	10·876	9·207	7·982	7·045	6·305	5·705	5·210	4·794	4·439	4·133
32	16·981	13·235	10·843	9·184	7·965	7·031	6·294	5·696	5·202	4·787	4·433	4·128
33	16·901	13·187	10·811	9·160	7·947	7·018	6·283	5·687	5·195	4·781	4·428	4·124
34	16·822	13·139	10·778	9·137	7·930	7·004	6·272	6·678	5·187	4·775	4·423	4·119
35	16·744	13·091	10·746	9·114	7·912	6·990	6·261	5·669	5·180	4·768	4·417	4·114
36	16·667	13·043	10·714	9·091	7·895	6·977	6·250	5·660	5·172	4·762	4·412	4·110
37	16·590	12·996	10·682	9·068	7·877	6·963	6·239	5·651	5·165	4·756	4·406	4·105
38	16·514	12·950	10·651	9·045	7·860	6·950	6·228	5·643	5·158	4·749	4·401	4·100
39	16·438	12·903	10·619	9·023	7·843	6·936	6·218	5·634	5·150	4·743	4·396	4·096

TABLE OF SPEEDS.

KNOTS PER HOUR.

Sec.	3	4	5	6	7	8	9	10	11	12	13	14
	Minutes.											
40	16·364	12·857	10·588	9·000	7·826	6·923	6·207	5·625	5·143	4·737	4·390	4·090
41	16·290	12·811	10·557	8·978	7·809	6·910	6·196	5·616	5·136	4·731	4·385	4·086
42	16·216	12·766	10·526	8·955	7·792	6·897	6·186	5·607	5·128	4·724	4·380	4·082
43	16·143	12·721	10·496	8·933	7·775	6·883	6·175	5·599	5·121	4·718	4·374	4·077
44	16·071	12·676	10·465	8·911	7·759	6·870	6·164	5·590	5·114	4·712	4·369	4·072
45	16·000	12·632	10·435	8·889	7·742	6·857	6·154	5·581	5·106	4·706	4·364	4·068
46	15·929	12·587	10·405	8·867	7·725	6·844	6·143	5·573	5·099	4·700	4·358	4·063
47	15·859	12·544	10·375	8·845	7·709	6·831	6·133	5·564	5·092	4·693	4·353	4·059
48	15·789	12·500	10·345	8·824	7·692	6·818	6·122	5·556	5·085	4·688	4·348	4·054
49	15·721	12·457	10·315	8·802	7·676	6·805	6·112	5·547	5·078	4·681	4·343	4·049

TABLE OF SPEEDS

KNOTS PER HOUR.

Sec.	3	4	5	6	7	8	9	10	11	12	13	14
	Minutes.											
50	15·652	12·414	10·286	8·780	7·660	6·792	6·102	5·538	5·070	4·675	4·337	4·045
51	15·584	12·371	10·256	8·759	7·643	6·780	6·091	5·530	5·063	4·669	4·332	4·040
52	15·517	12·329	10·227	8·738	7·627	6·767	6·081	5·521	5·056	4·663	4·327	4·035
53	15·451	12·287	10·198	8·717	7·611	6·754	6·071	5·513	5·049	4·657	4·322	4·031
54	15·385	12·245	10·169	8·696	7·595	6·742	6·061	5·505	5·042	4·651	4·317	4·027
55	15·319	12·203	10·141	8·675	7·579	6·729	6·050	5·496	5·035	4·645	4·311	4·022
56	15·254	12·162	10·112	8 654	7·563	6·716	6·040	5·488	5·028	4·639	4·306	4·018
57	15·190	12·121	10·084	8·633	7·547	6·704	6·030	5·479	5·021	4·633	4·301	4·013
58	15·126	12·081	10·056	8·612	7·531	6·691	6·020	5·471	5·014	4·627	4·296	4·009
59	15·063	12·040	10·028	8·592	7·516	6·679	6·010	5·463	5·007	4·621	4·291	4·004

A Table of Diameters, Areas, and Circumferences of Circles, from $\frac{1}{16}$ of an Inch to 110 Inches.

Dia.	Area.	Circum.	Dia.	Area.	Circum.
$\frac{1}{16}$	·0030	·1963	$\frac{1}{2}$	1.7671	4.7124
$\frac{1}{8}$	·0122	·3927	$\frac{9}{16}$	1.9175	4.9087
$\frac{3}{16}$	·0276	·5890	$\frac{5}{8}$	2.0739	5.1051
$\frac{1}{4}$	·0490	·7854	$\frac{11}{16}$	2.2365	5.3014
$\frac{5}{16}$	·0767	·9817	$\frac{3}{4}$	2.4052	5.4978
$\frac{3}{8}$	·1104	1·1781	$\frac{13}{16}$	2.5801	5.6941
$\frac{7}{16}$	·1503	1·3744	$\frac{7}{8}$	2.7611	5.8905
$\frac{1}{2}$	·1963	1·5708	$\frac{15}{16}$	2.9483	6.0868
$\frac{9}{16}$	·2485	1·7671			
$\frac{5}{8}$	·3068	1·9635	2 in.	3.1416	6.2832
$\frac{11}{16}$	·3712	2·1598	$\frac{1}{16}$	3.3411	6.4795
$\frac{3}{4}$	·4417	2·3562	$\frac{1}{8}$	3.5465	6.6759
$\frac{13}{16}$	·5185	2·5525	$\frac{3}{16}$	3.7582	6.8722
$\frac{7}{8}$	·6013	2·7489	$\frac{1}{4}$	3.9760	7.0686
$\frac{15}{16}$	·6903	2·9452	$\frac{5}{16}$	4.2001	7.2649
			$\frac{3}{8}$	4.4302	7.4613
1 in.	·7854	3·1416	$\frac{7}{16}$	4.6664	7.6576
$\frac{1}{16}$	·8861	3·3379	$\frac{1}{2}$	4.9087	7.8540
$\frac{1}{8}$	·9940	3·5343	$\frac{9}{16}$	5.1573	8.0503
$\frac{3}{16}$	1·1075	3·7306	$\frac{5}{8}$	5.4119	8.2467
$\frac{1}{4}$	1·2271	3·9270	$\frac{11}{16}$	5.6727	8.4430
$\frac{5}{16}$	1·3529	4·1233	$\frac{3}{4}$	5.9395	8.6394
$\frac{3}{8}$	1·4848	4·3197	$\frac{13}{16}$	6.2126	8.8357
$\frac{7}{16}$	1·6229	4·5160	$\frac{7}{8}$	6.4918	9.0321

Diameters, Areas, and Circumferences.

Dia.	Area.	Circum.	Dia.	Area.	Circum.
15/16	6·7772	9·2284	3/8	15·0331	13·7445
			7/16	15·4657	13·9408
3 in.	7·0686	9·4248	1/2	15·9043	14·1372
1/16	7·3662	9·6211	9/16	16·3492	14·3335
1/8	7·6699	9·8175	5/8	16·8001	14·5299
3/16	7·9798	10·0138	11/16	17·2573	14·7262
1/4	8·2957	10·2102	3/4	17·7205	14·9226
5/16	8·6179	10·4065	13/16	18·1900	15·1189
3/8	8·9462	10·6029	7/8	18·6655	15·3153
7/16	9·2806	10·7992	15/16	19·1472	15·5716
1/2	9·6211	10·9956			
9/16	9·9678	11·1919	5 in.	19·6350	15·7080
5/8	10·3206	11·3883	1/16	20·1290	15·9043
11/16	10·6796	11·5846	1/8	20·6290	16·1007
3/4	11·0446	11·7810	3/16	21·1252	16·2970
13/16	11·4159	11·9773	1/4	21·6475	16·4934
7/8	11·7932	12·1737	5/16	22·1661	16·6897
15/16	12·1768	12·3700	3/8	22·6907	16·8861
			7/16	23·2215	17·0824
4 in.	12·5664	12·5664	1/2	23·7583	17·2788
1/16	12·9622	12·7627	9/16	24·3014	17·4751
1/8	13·3640	12·9591	5/8	24·8505	17·6715
3/16	13·7721	13·1554	11/16	25·4058	17·8678
1/4	14·1862	13·3518	3/4	25·9672	18·0642
5/16	14·6066	13·5481	13/16	26·5348	18·2605

Diameters, Areas, and Circumferences.

Dia.	Area.	Circum.	Dia.	Area.	Circum.
$\frac{7}{8}$	27·1085	18·4569	$\frac{5}{16}$	41·9974	22·9729
$\frac{15}{16}$	27·6884	18·6532	$\frac{3}{8}$	42·7184	23·1693
			$\frac{7}{16}$	43·4455	23·3656
6 in.	28·2744	18·8496	$\frac{1}{2}$	44·1787	23·5620
$\frac{1}{16}$	28·8665	19·0459	$\frac{9}{16}$	44·9181	23·7583
$\frac{1}{8}$	29·4647	19·2423	$\frac{5}{8}$	45·6636	23·9547
$\frac{3}{16}$	30·0798	19·4386	$\frac{11}{16}$	46·4153	24·1510
$\frac{1}{4}$	30·6796	19·6350	$\frac{3}{4}$	47·1730	24·3474
$\frac{5}{16}$	31·2964	19·8313	$\frac{13}{16}$	47·9370	24·5437
$\frac{3}{8}$	31·9192	20·0277	$\frac{7}{8}$	48·7070	24·7401
$\frac{7}{16}$	32·5481	20·2240	$\frac{15}{16}$	49·4833	24·9364
$\frac{1}{2}$	33·1831	20·4204			
$\frac{9}{16}$	33·8244	20·6167	8 in.	50·2656	25·1328
$\frac{5}{8}$	34·4717	20·8131	$\frac{1}{16}$	51·0541	25·3291
$\frac{11}{16}$	35·1252	21·0094	$\frac{1}{8}$	51·8486	25·5255
$\frac{3}{4}$	35·7847	21·2058	$\frac{3}{16}$	52·8994	25·7218
$\frac{13}{16}$	36·4505	21·4021	$\frac{1}{4}$	53·4562	25·9182
$\frac{7}{8}$	37·1224	21·5985	$\frac{5}{16}$	54·2748	26·1145
$\frac{15}{16}$	37·8005	21·7948	$\frac{3}{8}$	55·0885	26·3109
			$\frac{7}{16}$	55·9138	26·5072
7 in.	38·4846	21·9912	$\frac{1}{2}$	56·7451	26·7036
$\frac{1}{16}$	39·1749	22·1875	$\frac{9}{16}$	57·5887	26·8999
$\frac{1}{8}$	39·8713	22·3839	$\frac{5}{8}$	58·4264	27·0963
$\frac{3}{16}$	40·5469	22·5802	$\frac{11}{16}$	59·7762	27·2926
$\frac{1}{4}$	41·2825	22·7766	$\frac{3}{4}$	60·1321	27·4890

Diameters, Areas, and Circumferences.

Dia.	Area.	Circum.	Dia.	Area.	Circum.
13/16	60·9943	27·6853	1/4	82·5160	32·2014
7/8	61·8625	27·8817	5/16	83·5254	32·3977
15/16	62·7369	28·0780	3/8	84·5409	32·5941
			7/16	85·5626	32·7904
9 in.	63·6174	28·2744	1/2	86·5903	32·9868
1/16	64·5041	28·4707	9/16	87·6243	33·1831
1/8	65·3968	28·6671	5/8	88·6643	33·3795
3/16	66·2957	28·8634	11/16	89·7105	33·5758
1/4	67·2007	29·0598	3/4	90·7627	33·7722
5/16	68·1120	29·2561	13/16	91·8212	33·9685
3/8	69·0293	29·4525	7/8	92·8858	34·1649
7/16	69·9528	29·6488	15/16	93·9566	34·3612
1/2	70·8823	29·8452			
9/16	71·8181	30·0415	11 in.	95·0334	34·5576
5/8	72·7599	30·2379	1/16	96·1164	34·7539
11/16	73·7079	30·4342	1/8	97·2053	34·9503
3/4	74·6620	30·6306	3/16	98·3008	35·1466
13/16	75·6223	30·8269	1/4	99·4021	35·3430
7/8	76·5887	31·0233	5/16	100·5097	35·5393
15/16	77·5613	31·2196	3/8	101·6234	35·7357
			7/16	102·7432	35·9320
10 in.	78·5400	31·4160	1/2	103·8691	36·1284
1/16	79·5248	31·6123	9/16	105·0012	36·3247
1/8	80·5157	31·8087	5/8	106·1394	36·5211
3/16	81·5128	32·0050	11/16	107·2838	36·7174

Diameters, Areas, and Circumferences.

Dia.	Area.	Circum.	Dia.	Area.	Circum.
¾	108·4342	36·9138	3/16	136·5890	41·4298
13/16	109·5909	37·1101	¼	137·8867	41·6262
⅞	110·7536	37·3065	5/16	139·1907	41·8225
15/16	111·9226	37·5028	⅜	140·5007	42·0189
			7/16	141·8169	42·2152
12 in.	113·0976	37·6992	½	143·1391	42·4116
1/16	114·2788	37·8955	9/16	144·4726	42·6079
⅛	115·4660	38·0919	⅝	145·8021	42·8043
3/16	116·6645	38·2882	11/16	147·1428	43·0006
¼	117·8590	38·4846	¾	148·4896	43·1970
5/16	119·0648	38·6809	13/16	149·8426	43·3933
⅜	120·2766	38·8773	⅞	151·2017	43·5897
7/16	121·4946	39·0736	15/16	152·5670	43·7860
½	122·7187	39·2700			
9/16	123·9490	39·4663	14 in.	153·9384	43·9824
⅝	125·1854	39·6627	1/16	155·3159	44·1787
11/16	126·4479	39·8590	⅛	156·6995	44·3751
¾	127·6765	40·0554	3/16	158·0893	44·5714
13/16	128·8999	40·2517	¼	159·4852	44·7676
⅞	130·1923	40·4481	5/16	160·8374	44·9641
15/16	131·4279	40·6444	⅜	162·2956	45·1605
			7/16	163·7099	45·3568
13 in.	132·7326	40·8408	½	165·1303	45·5532
1/16	134·0120	41·0371	9/16	166·5569	45·7495
⅛	135·2974	41·2338	⅝	167·9896	45·9459

Diameters, Areas, and Circumferences.

Dia.	Area.	Circum.	Dia.	Area.	Circum.
$\frac{11}{16}$	169·4285	46·1422	$\frac{1}{8}$	204·2162	50·6583
$\frac{3}{4}$	170·8735	46·3386	$\frac{3}{16}$	205·8024	50·8546
$\frac{13}{16}$	172·3247	46·5349	$\frac{1}{4}$	207·3946	51·0510
$\frac{7}{8}$	173·7820	46·7313	$\frac{5}{16}$	208·9931	51·2473
$\frac{15}{16}$	175·2455	46·9276	$\frac{3}{8}$	210·5976	51·4437
			$\frac{7}{16}$	212·2083	51·6400
15 in.	176·7150	47·1240	$\frac{1}{2}$	213·8251	51·8364
$\frac{1}{16}$	178·1907	47·3203	$\frac{9}{16}$	215·4481	52·0327
$\frac{1}{8}$	179·6725	47·5167	$\frac{5}{8}$	217·0772	52·2291
$\frac{3}{16}$	181·1105	47·7130	$\frac{11}{16}$	218 7124	52·4254
$\frac{1}{4}$	182·6545	47·9094	$\frac{3}{4}$	220·3537	52·6218
$\frac{5}{16}$	184·1548	48·1057	$\frac{13}{16}$	222·0013	52·8181
$\frac{3}{8}$	185·6612	48·3021	$\frac{7}{8}$	223·6549	53·0145
$\frac{7}{16}$	187·1737	48·4984	$\frac{15}{16}$	225·3147	53·2108
$\frac{1}{2}$	188·6923	48·6948			
$\frac{9}{16}$	190·2171	48·8911	17 in.	226·9806	53·4072
$\frac{5}{8}$	191·7480	49·0875	$\frac{1}{16}$	228·6527	53·6035
$\frac{11}{16}$	193·3351	49·2838	$\frac{1}{8}$	230·3308	53·7999
$\frac{3}{4}$	194·8282	49·4802	$\frac{3}{16}$	232·0151	53·9962
$\frac{13}{16}$	196·3776	49·6765	$\frac{1}{4}$	233·7055	54·1926
$\frac{7}{8}$	197·9330	49·8729	$\frac{5}{16}$	235·4022	54·3889
$\frac{15}{16}$	199·4947	50·0692	$\frac{3}{8}$	237·1049	54·5853
			$\frac{7}{16}$	238·8138	54·7816
16 in.	201·0624	50·2656	$\frac{1}{2}$	240·5287	54·9780
$\frac{1}{16}$	202·6363	50·4619	$\frac{9}{16}$	242·2499	55·1743

DIAMETERS, AREAS, AND CIRCUMFERENCES.

Dia.	Area.	Circum.	Dia.	Area.	Circum.
5/8	243·9771	55·3707	1/16	285·3978	59·8867
11/16	245·7105	55·5670	1/8	287·2723	60·0831
3/4	247·4500	55·7634	3/16	289·4030	60·2794
13/16	249·1952	55·9597	1/4	291·0397	60·4758
7/8	250·9475	56·1561	5/16	292·9324	60·6721
15/16	252·7050	56·3524	3/8	294·8312	60·8685
			7/16	296·7367	61·0648
18 in.	254·4696	56·5488	1/2	298·6483	61·2612
1/16	256·2398	56·7451	9/16	300·5658	61·4575
1/8	258·0161	56·9415	5/8	302·4894	61·6539
3/16	259·7986	57·1378	11/16	304·4192	61·8502
1/4	261·5872	57·3342	3/4	306·3550	62·0466
5/16	263·3820	57·5305	13/16	308·2971	62·2429
3/8	265·1829	57·7269	7/8	310·2452	62·4393
7/16	266·9900	57·9282	15/16	312·1996	62·6356
1/2	268·8031	58·1196			
9/16	270·6225	58·2159	20 in.	314·1600	62·8320
5/8	272·4479	58·5123	1/16	316·1266	63·0283
11/16	274·2895	58·7806	1/8	318·0992	63·2247
3/4	276·1171	58·9056	3/16	320·0781	63·4210
13/16	277·9610	59·1013	1/4	322·0630	63·6174
7/8	279·8110	59·2977	5/16	324·0542	63·8137
15/16	281·1672	58·4940	3/8	326·0514	64·0101
			7/16	328·0548	64·2064
19 in.	283·5294	59·6904	1/2	330·0643	64·4028

Diameters, Areas, and Circumferences.

Dia.	Area.	Circum.
9/16	332·0800	64·5991
5/8	334·1018	64·7955
11/16	336·1297	64·9918
3/4	338·1637	65·1882
13/16	340·2040	65·3845
7/8	342·2502	65·5809
15/16	344·3028	65·7772
21 in.	346·3614	65·7936
1/16	348·4267	66·1699
1/8	350·4970	66·3663
3/16	352·5740	66·5626
1/4	354·6571	66·7590
5/16	356·7465	66·9553
3/8	358·8419	67·1517
7/16	360·9435	67·3480
1/2	363·0511	67·5444
9/16	365·1650	67·7407
5/8	367·2849	67·9371
11/16	369·4110	68·1334
3/4	371·5432	68·3298
13/16	373·6816	68·5261
7/8	375·8261	68·7225
15/16	377·9768	68·9188
22 in.	380·1336	69·1152
1/16	382·2965	69·3115
1/8	384·4655	69·5079
3/16	386·6907	69·7042
1/4	388·8220	69·9006
5/16	391·0095	70·0969
3/8	393·2031	70·2933
7/16	395·4039	70·4806
1/2	397·6087	70·6860
9/16	399·8207	70·8823
5/8	402·0388	71·0787
11/16	404·2631	71·2750
3/4	406·4935	71·4714
13/16	408·7301	71·6677
7/8	410·9728	71·8641
15/16	413·2317	72·0604
23 in.	415·4766	72·2568
1/16	417·7377	72·4531
1/8	420·0049	72·6495
3/16	422·2783	72·8458
1/4	424·5577	73·0422
5/16	426·3434	73·2385
3/8	429·1352	73·4349
7/16	431·4331	73·6312

Diameters, Areas, and Circumferences.

Dia.	Area.	Circum.	Dia.	Area.	Circum.
1/2	433·7371	73·8276	25 in.	490·8750	78·5400
9/16	436·0473	74·0239	1/16	493·3325	78·7363
5/8	438·3636	74·2203	1/8	495·7960	78·9327
11/16	440·6811	74·4166	3/16	498·2657	79·1290
3/4	443·0146	74·6130	1/4	500·7415	79·3254
13/16	445·3539	74·8093	5/16	503·2236	79·5217
7/8	447·6992	75·0057	3/8	505·7117	79·7181
15/16	450·0418	75·2020	7/16	508·2060	79·9144
			1/2	510·7063	80·1108
24 in.	452·3904	75·3984	9/16	513·2129	80·3071
1/16	454·7497	75·5947	5/8	515·7255	80·5035
1/8	457·1150	75·7911	11/16	518·2443	80·6998
3/16	459·4866	75·9874	3/4	520·7692	80·8962
1/4	461·8642	76·1838	13/16	523·3003	81·0925
5/16	464·2481	76·3801	7/8	525·8375	81·2889
3/8	466·6380	76·5765	15/16	528·3809	81·4852
7/16	469·0341	76·7728			
1/2	471·4363	76·9692	26 in.	530·9304	81·6816
9/16	473·8447	77·1655	1/16	533·4860	81·8779
5/8	476·2592	77·3619	1/8	536·0477	82·0743
11/16	478·6798	77·5582	3/16	538·6156	82·2706
3/4	481·1065	77·7546	1/4	541·1896	82·4670
13/16	483·5395	77·9509	5/16	543·7698	82·6633
7/8	485·9785	78·1473	3/8	546·3561	82·8597
15/16	488·4237	78·3436	7/16	548·9486	83·0560

Diameters, Areas, and Circumferences.

Dia.	Area.	Circum.	Dia.	Area.	Circum.
1/2	551·5471	83·2524	28 in.	615·7536	87·9648
9/16	554·1519	83·4487	1/16	618·5051	88·1611
5/8	556·7627	83·6451	1/8	621·2636	88·3574
11/16	559·3797	83·8414	3/16	624·0279	88·5538
3/4	562·0027	84·0378	1/4	626·7982	88·7502
13/16	564·6320	84·2341	5/16	629·5748	88·9465
7/8	567·2674	84·4305	3/8	632·3574	89·1429
15/16	569·4090	84·6268	7/16	635·1462	89·3392
			1/2	637·9411	89·5356
27 in.	572·5566	84·8232	9/16	640·7422	89·7319
1/16	575·2104	85·0195	5/8	643·5494	89·9283
1/8	577·8703	85·2159	11/16	646·3627	90·1246
3/16	580·5364	85·4122	3/4	649·1821	90·3210
1/4	583·2085	85·6086	13/16	652·0078	90·5173
5/16	585·8869	85·8049	7/8	654·8395	90·7137
3/8	588·5714	86·0013	15/16	657·6774	90·9100
7/16	591·2620	86·1976			
1/2	593·9587	86·3940	29 in.	660·5214	91·1064
9/16	596·6616	86·5903	1/16	663·3716	91·3027
5/8	599·3706	86·7867	1/8	666·2278	91·4991
11/16	602·0858	86·9830	3/16	669·0902	91·6954
3/4	604·8070	87·1794	1/4	671·9587	91·8918
13/16	607·5345	87·3757	5/16	674·8335	92·0081
7/8	610·2680	87·5721	3/8	677·7143	92·2845
15/16	613·0078	87·7684	7/16	680·6013	92·4808

Diameters, Areas, and Circumferences.

Dia.	Area.	Circnm.	Dia.	Area.	Circum.
$\frac{1}{2}$	683·4943	92·6772	31 in.	754·7694	97·3896
$\frac{9}{16}$	686·3936	92·8735	$\frac{1}{16}$	757·8159	97·5859
$\frac{5}{8}$	689·2989	93·0699	$\frac{1}{8}$	760·8685	97·7823
$\frac{11}{16}$	692·2104	93·2662	$\frac{3}{16}$	763·9273	97·9786
$\frac{3}{4}$	695·1280	93·4626	$\frac{1}{4}$	766·9921	98·1750
$\frac{13}{16}$	698·0518	93·6589	$\frac{5}{16}$	770·0632	98·3713
$\frac{7}{8}$	700·9817	93·8553	$\frac{3}{8}$	773·1404	98·5677
$\frac{15}{16}$	703·9178	94·0516	$\frac{7}{16}$	776·2237	98·7648
			$\frac{1}{2}$	779·3131	98·9684
30 in.	706·8600	94·2480	$\frac{9}{16}$	782·4087	99·1567
$\frac{1}{16}$	709·8083	94·4443	$\frac{5}{8}$	785·5104	99·3531
$\frac{1}{8}$	712·7627	94·6407	$\frac{11}{16}$	788·6183	99·5494
$\frac{3}{16}$	715·7233	94·8370	$\frac{3}{4}$	791·7322	99·7458
$\frac{1}{4}$	718·6900	95·0334	$\frac{13}{16}$	794·8524	99·9421
$\frac{5}{16}$	721·6629	95·2297	$\frac{7}{8}$	797·9786	100·1385
$\frac{3}{8}$	724·6419	95·4261	$\frac{15}{16}$	801·1111	100·3348
$\frac{7}{16}$	727·6271	95·6224			
$\frac{1}{2}$	730·6183	95·8188	32 in.	804·2496	100·5312
$\frac{9}{16}$	733·6158	96·0151	$\frac{1}{16}$	807·3943	100·7275
$\frac{5}{8}$	736·6193	96·2115	$\frac{1}{8}$	810·5453	100·9240
$\frac{11}{16}$	739·6290	96·4078	$\frac{3}{16}$	813·7020	101·1202
$\frac{3}{4}$	742·6447	96·6042	$\frac{1}{4}$	816·8650	101·3166
$\frac{13}{16}$	745·6667	96·8005	$\frac{5}{16}$	820·0343	101·5130
$\frac{7}{8}$	748·6948	96·9969	$\frac{3}{8}$	823·2096	101·7093
$\frac{15}{16}$	751·7291	97·1932	$\frac{7}{16}$	826·3911	101·9056

Diameters, Areas, and Circumferences.

Dia.	Area.	Circum.	Dia.	Area.	Circum.
1/2	829·5787	102·1020	34 in.	907·9224	106·8144
9/16	832·7725	102·2983	1/16	911·2645	107·0107
5/8	835·9724	102·4947	1/8	914·6105	107·2071
11/16	839·1784	102·6910	3/16	917·9640	107·4034
3/4	842·3905	102·8874	1/4	921·3232	107·5998
13/16	845·6089	103·0837	5/16	924·6883	107·7961
7/8	848·8333	103·2801	3/8	928·0605	107·9925
15/16	852·0639	103·4764	7/16	931·4380	108·1888
			1/2	934·8223	108·3852
33 in.	855·3006	103·6728	9/16	938·2121	108·5815
1/16	858·5436	103·8691	5/8	941·6087	108·7779
1/8	861·7924	104·0655	11/16	945·0110	108·9742
3/16	865·0475	104·2618	3/4	948·4195	109·1706
1/4	868·3087	104·4582	13/16	951·8341	109·3669
5/16	871·5760	104·6545	7/8	955·2550	109·5633
3/8	874·8497	104·8509	15/16	958·6820	109·7596
7/16	878·1290	105·0472			
1/2	881·4151	105·2436	35 in.	962·1150	109·9560
9/16	884·7070	105·4399	1/16	965·5542	110·1523
5/8	888·0051	105·6363	1/8	968·9995	110·3487
11/16	891·3090	105·8326	3/16	972·4510	110·5450
3/4	894·6196	106·0290	1/4	975·9085	110·7414
13/16	897·9369	106·2253	5/16	979·3686	110·9377
7/8	901·2587	106·4217	3/8	982·8422	111·1341
15/16	904·5875	106·6180	7/16	986·3180	111·3304

Diameters, Areas, and Circumferences.

Dia.	Area.	Circum.	Dia.	Area.	Circum.
1/2	989·8003	111·5268	37 in.	1075·2126	116·2392
9/16	993·2097	111·7231	1/16	1078·8482	116·4355
5/8	996·7830	111·9195	1/8	1082·4898	116·6319
11/16	1000·3472	112·1158	3/16	1086·1376	116·8282
3/4	1003·7902	112·3122	1/4	1089·7915	117·0246
13/16	1007·3030	112·5086	5/16	1093·4517	117·2209
7/8	1010·8220	112·7049	3/8	1097·1179	117·4173
15/16	1014·3472	112·9012	7/16	1100·7903	117·6136
			1/2	1104·4687	117·8100
36 in.	1017·8784	113·0976	9/16	1108·1534	118·0063
1/16	1021·4158	113·2939	5/8	1111·8441	118·2027
1/8	1024·9592	113·4903	11/16	1115·5410	118·3990
3/16	1028·5089	113·6866	3/4	1119·2440	118·5954
1/4	1032·0646	113·8830	13/16	1122·9532	118·7917
5/16	1035·6266	114·0793	7/8	1126·6685	118·9881
3/8	1039·1946	114·2757	15/16	1130·3900	119·1844
7/16	1042·7913	114·4720			
1/2	1046·3941	114·6684	38 in.	1134·1176	119·3808
9/16	1049·9581	114·8647	1/16	1137·8513	119·5771
5/8	1053·5281	115·0611	1/8	1141·5911	119·7735
11/16	1057·1269	115·2572	3/16	1145·3371	119·9698
3/4	1060·7317	115·4538	1/4	1149·0892	120·1662
13/16	1064·3428	115·6501	5/16	1152·8475	120·3625
7/8	1067·9599	115·8465	3/8	1156·6119	120·5589
15/16	1071·5832	116·0428	7/16	1160·3825	120·7552

Diameters, Areas, and Circumferences.

Dia.	Area.	Circum.	Dia.	Area.	Circum.
½	1164·1591	120·9516	40 in.	1256·6400	125·6640
9/16	1167·9420	121·1479	1/16	1260·5701	125·8603
5/8	1171·7309	121·3443	⅛	1264·5062	126·0567
11/16	1175·5260	121·5406	3/16	1268·4486	126·2530
¾	1179·3271	121·7370	¼	1272·3970	126·4494
13/16	1183·1345	121·9333	5/16	1276·3517	126·6457
⅞	1186·9480	122·1297	⅜	1280·3124	126·8421
15/16	1190·7677	122·3260	7/16	1284·2793	127·0384
			½	1288·2523	127·2348
39 in.	1194·5934	122·5224	9/16	1292·2315	127·4311
1/16	1198·4253	122·7187	⅝	1296·2168	127·6275
⅛	1202·2633	122·9151	11/16	1300·2082	127·8238
3/16	1206·1075	123·1114	¾	1304·2057	128·0202
¼	1209·9577	123·3078	13/16	1308·2095	128·2165
5/16	1213·8142	123·5041	⅞	1312·2193	128·4129
⅜	1217·6768	123·7005	15/16	1316·2353	128·6092
7/16	1221·5455	123·8968			
½	1225·4203	124·0932	41 in.	1320·2574	128·8056
9/16	1229·3013	124·2895	1/16	1324·2857	129·0019
⅝	1233·1884	124·4859	⅛	1328·3200	129·1983
11/16	1237·0817	124·6822	3/16	1332·3605	129·3946
¾	1240·9810	124·8786	¼	1336·4071	129·5910
13/16	1244·8866	125·0749	5/16	1340·4600	129·7873
⅞	1248·7982	125·2713	⅜	1344·5189	129·9837
15/16	1252·7161	125·4676	7/16	1348·5840	130·1800

Diameters, Areas, and Circumferences.

Dia.	Area.	Circum.	Dia.	Area.	Circum.
1/2	1352·6551	130·3764	43 in.	1452·2046	135·0888
9/16	1356·7325	130·5727	1/16	1456·4292	135·2851
5/8	1360·8159	130·7691	1/8	1460·6599	135·4815
11/16	1364·9055	130·9654	3/16	1464·8968	135·6778
3/4	1369·0012	131·1618	1/4	1469·1397	135·8742
13/16	1373·1031	131·3581	5/16	1473·3839	136·0705
7/8	1377·2111	131·5545	3/8	1477·6342	136·2669
15/16	1381·3253	131·7508	7/16	1481·9006	136·4632
			1/2	1486·1731	136·6596
42 in.	1385·4456	131·9472	9/16	1490·4468	136·8559
1/16	1389·5720	132·1435	5/8	1494·7266	137·0523
1/8	1393·7045	132·3399	11/16	1499·0126	137·2486
3/16	1397·8432	132·5362	3/4	1503·3046	137·4450
1/4	1401·9880	132·7326	13/16	1507·6029	137·6413
5/16	1406·1390	132·9289	7/8	1511·9072	137·8377
3/8	1410·2961	133·1253	15/16	1516·2178	138·0340
7/16	1414·4594	133·3216			
1/2	1418·6287	133·5180	44 in.	1520·5344	138·2304
9/16	1422·8043	133·7143	1/16	1524·8572	138·4267
5/8	1426·9859	133·9107	1/8	1529·1860	138·6231
11/16	1431·1737	134·1070	3/16	1533·5211	138·8194
3/4	1435·3675	134·3034	1/4	1537·8622	139·0158
13/16	1439·5676	134·4997	5/16	1542·2046	139·2121
7/8	1443·7738	134·6961	3/8	1546·5530	139·4085
15/16	1447·9862	134·8924	7/16	1550·9176	139·6048

Diameters, Areas, and Circumferences.

Dia.	Area.	Circum.	Dia.	Area.	Circum.
1/2	1555·2883	139·8012	46 in.	1661·9064	144·5136
9/16	1559·6602	139·9975	1/16	1666·4255	144·7099
5/8	1564·0382	140·1939	1/8	1670·9507	144·9063
11/16	1568·4223	140·3902	3/16	1675·4821	145·1026
3/4	1572·8125	140·5866	1/4	1680·0196	145·2990
13/16	1577·2090	140·7829	5/16	1684·5583	145·4953
7/8	1581·6115	140·9793	3/8	1689·1031	145·6917
15/16	1586·0203	141·1756	7/16	1693·6641	145·8880
			1/2	1698·2311	146·0844
45 in.	1590·4350	141·3720	9/16	1702·7994	146·2807
1/16	1594·4560	141·5683	5/8	1707·3737	146·4771
1/8	1599·2830	141·7647	11/16	1711·9542	146·6734
3/16	1603·7162	141·9610	3/4	1716·5407	149·8698
1/4	1608·1555	142·1574	13/16	1721·1335	147·0661
5/16	1612·5961	142·3537	7/8	1725·7324	147·2625
3/8	1617·0427	142·5501	15/16	1730·3375	147·4588
7/16	1621·5055	142·7464			
1/2	1625·9743	142·9428	47 in.	1734·9486	147·6552
9/16	1630·4444	143·1391	1/16	1739·5659	147·8515
5/8	1634·9205	143·3355	1/8	1744·1893	148·0479
11/16	1639·4028	143·5318	3/16	1748·8189	148·2442
3/4	1643·8912	143·7282	1/4	1753·4545	148·4406
13/16	1648·3858	143·9245	5/16	1758·0914	148·6369
7/8	1652·8865	144·1209	3/8	1762·7344	148·8333
15/16	1657·3934	144·3172	7/16	1767·3935	149.0296

Diameters, Areas, and Circumferences.

Dia.	Area.	Circum.	Dia.	Area.	Circum.
$\frac{1}{2}$	1772·0587	149·2260	49 in.	1885·7454	153·9384
$\frac{9}{16}$	1776·7251	149·4223	$\frac{1}{16}$	1890·5591	154·1347
$\frac{5}{8}$	1781·3976	149·6187	$\frac{1}{8}$	1895·3788	154·3311
$\frac{11}{16}$	1786·0763	149·8150	$\frac{3}{16}$	1900·2047	154·5274
$\frac{3}{4}$	1790·7610	150·0114	$\frac{1}{4}$	1905·0367	154·7238
$\frac{13}{16}$	1795·4520	150·2077	$\frac{5}{16}$	1909·8700	154·9201
$\frac{7}{8}$	1800·1490	150·4041	$\frac{3}{8}$	1914·7093	155·1165
$\frac{15}{16}$	1804·8523	150·6004	$\frac{7}{16}$	1919·5648	155·3128
			$\frac{1}{2}$	1924·4263	155·5092
48 in.	1809·5616	150·7968	$\frac{9}{16}$	1929·2891	155·7055
$\frac{1}{16}$	1814·2551	150·9931	$\frac{5}{8}$	1934·1579	155·9019
$\frac{1}{8}$	1818·9986	151·1895	$\frac{11}{16}$	1939·0329	156·0982
$\frac{3}{16}$	1823·7264	151·3858	$\frac{3}{4}$	1943·9140	156·2946
$\frac{1}{4}$	1828·4602	151 5822	$\frac{13}{16}$	1948·8013	156·4909
$\frac{5}{16}$	1833·1953	151·7785	$\frac{7}{8}$	1953·6947	156·6873
$\frac{3}{8}$	1837·9364	151·9749	$\frac{15}{16}$	1958·0943	156·8836
$\frac{7}{16}$	1842·6937	152·1712			
$\frac{1}{2}$	1847·4571	152·3676	50 in.	1963·5000	157·0800
$\frac{9}{16}$	1852·2167	152·5639	$\frac{1}{16}$	1968·4118	157·2763
$\frac{5}{8}$	1856·9924	152·7603	$\frac{1}{8}$	1973·3297	157·4727
$\frac{11}{16}$	1861·7892	152·9566	$\frac{3}{16}$	1978·2525	157·6690
$\frac{3}{4}$	1868·5521	153·1530	$\frac{1}{4}$	1983·1840	157·8654
$\frac{13}{16}$	1871·3413	153·3493	$\frac{5}{16}$	1988·6154	158·0617
$\frac{7}{8}$	1876·1365	153·5457	$\frac{3}{8}$	1993·0529	158·2581
$\frac{15}{16}$	1880·9379	153·7420	$\frac{7}{16}$	1998·0066	158·4544

Diameters, Areas, and Circumferences.

Dia.	Area.	Circum.	Dia.	Area.	Circum.
$\frac{1}{2}$	2002·9663	158·6508	52 in.	2123·7216	163·3632
$\frac{9}{16}$	2007·9273	158·8471	$\frac{1}{16}$	2128·8298	163·5595
$\frac{5}{8}$	2012·8943	159·0435	$\frac{1}{8}$	2133·9440	163·7559
$\frac{11}{16}$	2017·8675	159·2398	$\frac{3}{16}$	2139·0645	163·9522
$\frac{3}{4}$	2022·8467	159·4362	$\frac{1}{4}$	2144·1910	164·1486
$\frac{13}{16}$	2027·8172	159·6325	$\frac{5}{16}$	2149·3238	164·3449
$\frac{7}{8}$	2032·8238	159·8289	$\frac{3}{8}$	2154·4626	164·5413
$\frac{15}{16}$	2037·8216	160·0252	$\frac{7}{16}$	2159·6076	164·7376
			$\frac{1}{2}$	2164·7587	164·9340
51 in.	2042·8254	160·2216	$\frac{9}{16}$	2169·9160	165·1303
$\frac{1}{16}$	2047·8354	160·4179	$\frac{5}{8}$	2175·0794	165·3267
$\frac{1}{8}$	2052·8515	160·6143	$\frac{11}{16}$	2180·2489	165·5230
$\frac{3}{16}$	2057·8798	160·8106	$\frac{3}{4}$	2185·4245	165·7194
$\frac{1}{4}$	2062·9021	161·0070	$\frac{13}{16}$	2190·6064	165·9157
$\frac{5}{16}$	2067·9317	161·2033	$\frac{7}{8}$	2195·7943	166·1121
$\frac{3}{8}$	2072·9674	161·3997	$\frac{15}{16}$	2200·9884	166·3084
$\frac{7}{16}$	2078·0293	161·5960			
$\frac{1}{2}$	2083·0771	161·7924	53 in.	2206·1886	166·5048
$\frac{9}{16}$	2088·1362	161·9887	$\frac{1}{16}$	2211·3950	166·7011
$\frac{5}{8}$	2093·2014	162·1851	$\frac{1}{8}$	2216·6074	166·8975
$\frac{11}{16}$	2098·2678	162·3814	$\frac{3}{16}$	2221·8260	167·0938
$\frac{3}{4}$	2103·3502	162·5778	$\frac{1}{4}$	2227·0507	167·2902
$\frac{13}{16}$	2108·4339	162·7741	$\frac{5}{16}$	2232·2817	167·4865
$\frac{7}{8}$	2113·5236	162·9705	$\frac{3}{8}$	2237·5187	167·6829
$\frac{15}{16}$	2118·1196	163·1668	$\frac{7}{16}$	2242·7619	167·8792

Diameters, Areas, and Circumferences.

Dia.	Area.	Circum.	Dia.	Area.	Circum.
$\frac{1}{2}$	2248.0111	168·0756	55 in.	2375·8350	172·7880
$\frac{9}{16}$	2253·2666	168·2718	$\frac{1}{16}$	2381·2382	172·9843
$\frac{5}{8}$	2258·5281	168·4683	$\frac{1}{8}$	2386·6465	173·1807
$\frac{11}{16}$	2263·7908	168·6646	$\frac{3}{16}$	2392·0515	173·3770
$\frac{3}{4}$	2269·0696	168·8610	$\frac{1}{4}$	2397·4825	173·5734
$\frac{13}{16}$	2274·3496	169·0573	$\frac{5}{16}$	2402·9098	173·7697
$\frac{7}{8}$	2279·6357	169·2537	$\frac{3}{8}$	2408·3432	173·9661
$\frac{15}{16}$	2284·9280	169·4500	$\frac{7}{16}$	2413·7777	174·1624
			$\frac{1}{2}$	2419·2283	174·3588
54 in.	2290·2264	169·6464	$\frac{9}{16}$	2424·7026	174·5551
$\frac{1}{16}$	2295·5309	169·8427	$\frac{5}{8}$	2430·1830	174·7515
$\frac{1}{8}$	2300·8415	170·0391	$\frac{11}{16}$	2435·6246	174·9478
$\frac{3}{16}$	2306·1583	170·2354	$\frac{3}{4}$	2441·0722	175·1442
$\frac{1}{4}$	2311·4812	170·4318	$\frac{13}{16}$	2446·5486	175·3405
$\frac{5}{16}$	2316·8163	170·6281	$\frac{7}{8}$	2452·0310	175·5369
$\frac{3}{8}$	2322·1455	170·8245	$\frac{15}{16}$	2457·0197	175·7332
$\frac{7}{16}$	2327·4819	171·0208			
$\frac{1}{2}$	2332·8343	171·2172	56 in.	2463·0144	175·9296
$\frac{9}{16}$	2338·1880	171·4135	$\frac{1}{16}$	2468·5153	176·1259
$\frac{5}{8}$	2343·5477	171·6099	$\frac{1}{8}$	2474·0222	176·3223
$\frac{11}{16}$	2348·9636	171·8062	$\frac{3}{16}$	2479·5354	176·5186
$\frac{3}{4}$	2354·2855	172·0026	$\frac{1}{4}$	2485·0546	176·7150
$\frac{13}{16}$	2359·6637	172·1989	$\frac{5}{16}$	2490·5351	176·9913
$\frac{7}{8}$	2365·0480	172·3953	$\frac{3}{8}$	2496·1116	177·1077
$\frac{15}{16}$	2370·4385	172·5916	$\frac{7}{16}$	2501·6493	177·3040

Diameters, Areas, and Circumferences.

Dia.	Area.	Circum.	Dia.	Area.	Circum.
$\frac{1}{2}$	2507·1931	177·5044	58 in.	2642·0856	182·2128
$\frac{9}{16}$	2512·7431	177.6967	$\frac{1}{16}$	2647·7328	182·4091
$\frac{5}{8}$	2518·2992	177.8931	$\frac{1}{8}$	2653·4851	182·6055
$\frac{11}{16}$	2523·8614	178·0894	$\frac{3}{16}$	2659·9565	182·8018
$\frac{3}{4}$	2529·4297	178·2858	$\frac{1}{4}$	2664·9112	182·9982
$\frac{13}{16}$	2535·0043	178·4821	$\frac{5}{16}$	2670·6330	183·1945
$\frac{7}{8}$	2540·5849	178·6785	$\frac{3}{8}$	2676·3609	183·3909
$\frac{15}{16}$	2546·1717	178·8748	$\frac{7}{16}$	2682·0950	183·5872
			$\frac{1}{2}$	2687·8351	183·7836
57 in.	2551·7646	179·0712	$\frac{9}{16}$	2693·5814	183·9799
$\frac{1}{16}$	2557·3637	179·2675	$\frac{5}{8}$	2699·3338	184·1763
$\frac{1}{8}$	2562·9688	179·4639	$\frac{11}{16}$	2705·0924	184·3726
$\frac{3}{16}$	2568·5801	179·6602	$\frac{3}{4}$	2710·8571	184·5690
$\frac{1}{4}$	2574·1975	179·8566	$\frac{13}{16}$	2716·6280	184·7653
$\frac{5}{16}$	2579·8212	180·0529	$\frac{7}{8}$	2722·4050	184·9617
$\frac{3}{8}$	2585·4509	180·2493	$\frac{15}{16}$	2728·1882	185·1580
$\frac{7}{16}$	2591·0869	180·4456			
$\frac{1}{2}$	2496·7287	180·6420	59 in.	2733·9774	185·3544
$\frac{9}{16}$	2602·3769	180·8383	$\frac{1}{16}$	2739·7728	185·5507
$\frac{5}{8}$	2608·0311	181·0347	$\frac{1}{8}$	2745·5743	185·7471
$\frac{11}{16}$	2613·6942	181·2310	$\frac{3}{16}$	2751·8820	185·9434
$\frac{3}{4}$	2619·3580	181·4274	$\frac{1}{4}$	2757·1957	186·1398
$\frac{13}{16}$	2625·0307	181·6237	$\frac{5}{16}$	2763·0157	186·3361
$\frac{7}{8}$	2630·7095	181·8201	$\frac{3}{8}$	2768·8418	186·5325
$\frac{15}{16}$	2636·3945	182·0164	$\frac{7}{16}$	2774·6745	186·7288

Diameters, Areas, and Circumferences.

Dia.	Area.	Circum.	Dia.	Area.	Circum.
$\frac{1}{2}$	2780·5123	186·9252	61 in.	2922·4734	191·6376
$\frac{9}{16}$	2786·3568	187·1215	$\frac{1}{16}$	2928·4652	191·8339
$\frac{5}{8}$	2792·2074	187·3179	$\frac{1}{8}$	2934·4630	192·0303
$\frac{11}{16}$	2798·0642	187·5142	$\frac{3}{16}$	2940·4670	192·2266
$\frac{3}{4}$	2803·9270	187·7106	$\frac{1}{4}$	2946·4771	192·4230
$\frac{13}{16}$	2809·7461	187·9069	$\frac{5}{16}$	2952·4938	192·6193
$\frac{7}{8}$	2815·6712	188·1033	$\frac{3}{8}$	2958·5159	192·8157
$\frac{15}{16}$	2821·5526	188·2996	$\frac{7}{16}$	2964·5445	193·0120
			$\frac{1}{2}$	2970·5791	193·2084
60 in.	2827·4400	188·4960	$\frac{9}{16}$	2976·6200	193·4047
$\frac{1}{16}$	2833·3336	188·6923	$\frac{5}{8}$	2982·6669	193·6011
$\frac{1}{8}$	2839·2332	188·8887	$\frac{11}{16}$	2988·7200	193·7974
$\frac{3}{16}$	2845·1391	189·0850	$\frac{3}{4}$	2994·7792	193·9938
$\frac{1}{4}$	2851·0510	189·2814	$\frac{13}{16}$	3000·8423	194·1901
$\frac{5}{16}$	2856·9692	189·4777	$\frac{7}{8}$	3006·9161	194·3865
$\frac{3}{8}$	2862·8934	189·6741	$\frac{15}{16}$	3017·9938	194·5828
$\frac{7}{16}$	2868·8223	189·8704			
$\frac{1}{2}$	2874·7603	189·0668	62 in.	3019·0776	194·7792
$\frac{9}{16}$	2880·7030	190·2631	$\frac{1}{16}$	3025·1675	194·9755
$\frac{5}{8}$	2886·6517	190·4595	$\frac{1}{8}$	3031·2635	195·1719
$\frac{11}{16}$	2892·6067	190·6558	$\frac{3}{16}$	3037·3607	195·3682
$\frac{3}{4}$	2898·5677	190·8522	$\frac{1}{4}$	3043·4740	195·5646
$\frac{13}{16}$	2904·5350	191·0485	$\frac{5}{16}$	3049·6885	195·7609
$\frac{7}{8}$	2910·5083	191·2449	$\frac{3}{8}$	3055·7091	195·9573
$\frac{15}{16}$	2916·4878	191·4412	$\frac{7}{16}$	3061·8359	196·1536

Diameters, Areas, and Circumferences.

Dia.	Area.	Circum.	Dia.	Area.	Circnm.
$\frac{1}{2}$	3067·9687	196·3500	64 in.	3216·9984	201·0624
$\frac{9}{16}$	3074·1578	196·5463	$\frac{1}{16}$	3223·2847	201·2587
$\frac{5}{8}$	3080·2529	196·7427	$\frac{1}{8}$	3229·5770	201·4551
$\frac{11}{16}$	3086·4042	196·9390	$\frac{3}{16}$	3235·8746	201·6514
$\frac{3}{4}$	3092·5615	197·1354	$\frac{1}{4}$	3242·1782	201·8478
$\frac{13}{16}$	3098·7251	197·3317	$\frac{5}{16}$	3248·4936	202·0441
$\frac{7}{8}$	3104·8948	197·5281	$\frac{3}{8}$	3254·8080	202·2405
$\frac{15}{16}$	3111·0707	197·7244	$\frac{7}{16}$	3261·1311	202·4368
			$\frac{1}{2}$	3267·4603	202·6332
63 in.	3117·2526	197·9208	$\frac{9}{16}$	3273·7957	202·8295
$\frac{1}{16}$	3124·4407	198·1171	$\frac{5}{8}$	3280·1372	203·0295
$\frac{1}{8}$	3129·6349	198·3135	$\frac{11}{16}$	3286·4875	203·2222
$\frac{3}{16}$	3135·8353	198·5098	$\frac{3}{4}$	3292·8385	203·4186
$\frac{1}{4}$	3142·0417	198·7062	$\frac{13}{16}$	3299·1985	203·6149
$\frac{5}{16}$	3148·7544	198·9025	$\frac{7}{8}$	3305·5645	203·8113
$\frac{3}{8}$	3154·4732	199·0989	$\frac{15}{16}$	3311·9367	204·0076
$\frac{7}{16}$	3160·7981	199·2952			
$\frac{1}{2}$	3166·9291	199·4916	65 in.	3318·3151	204·2040
$\frac{9}{16}$	3173·1663	199·6879	$\frac{1}{16}$	3324·7495	204·4003
$\frac{5}{8}$	3179·4096	199·8843	$\frac{1}{8}$	3331·0900	204·5917
$\frac{11}{16}$	3185·6591	200·0806	$\frac{3}{16}$	3337·9857	204·7930
$\frac{3}{4}$	3191·9146	200·2770	$\frac{1}{4}$	3343·8875	204·9894
$\frac{13}{16}$	3193·1764	200·4733	$\frac{5}{16}$	3350·2976	205·1857
$\frac{7}{8}$	3204·4442	200·6697	$\frac{3}{8}$	3356·7137	205·3821
$\frac{15}{16}$	3210·7183	200·8660	$\frac{7}{16}$	3363·1350	205·5784

Diameters, Areas, and Circumferences.

Dia.	Area.	Circum.	Dia.	Area.	Circum.
1/2	3369·2623	205·7748	67 in.	3525·6606	210·4872
9/16	3375·9959	205·9711	1/16	3532·2414	210·6835
5/8	3382·4355	206·1675	1/8	3538·8283	210·8799
11/16	3388·8813	206·3638	3/16	3545·4200	211·0762
3/4	3395·3332	206·5602	1/4	3552·0185	211·2726
13/16	3401·7913	206·7565	5/16	3558·6249	211·4689
7/8	3408·2555	206·9529	3/8	3565·2374	211·6653
15/16	3414·7259	207·1492	7/16	3571·8550	211·8616
			1/2	3578·4787	212·0580
66 in.	3421·2024	207·3456	9/16	3585·1086	212·2543
1/16	3427·6850	207·5491	5/8	3591·7446	212·4507
1/8	3434·1737	207·7383	11/16	3598·8868	212·6470
3/16	3440·6676	207·9346	3/4	3605·0350	212·8434
1/4	3447·1676	208·1310	13/16	3611·6895	213·0397
5/16	3453·6758	208·3273	7/8	3618·3500	213·2361
3/8	3468·1901	208·5237	15/16	3625·0168	213·4324
7/16	3470·7096	208·7200			
1/2	3473·2351	208·9164	68 in.	3631·6896	213·6288
9/16	3479·7669	209·1127	1/16	3638·3686	213·8251
5/8	3486·3047	209·3091	1/8	3645·0536	214·0215
11/16	3492·8487	209·5054	3/16	3651·7439	214·2178
3/4	3499·3987	209·7018	1/4	3658·4402	214·4142
13/16	3506·4550	209·8981	5/16	3665·1448	214·6105
7/8	3512·5174	210·0945	3/8	3671·8554	214·8069
15/16	3519·0860	210·2908	7/16	3678·5762	215·0032

Diameters, Areas, and Circumferences.

Dia.	Area.	Circum.	Dia.	Area.	Circum.
½	3685·2931	215·1996	70 in.	3848·4600	219·9120
9/16	3693·0212	215·3959	1/16	3855·8353	220·1083
⅝	3698·7554	215·5923	⅛	3862·2167	220·3047
11/16	3703·9957	215·7886	3/16	3869·1033	220·5010
¾	3712·2421	215·9850	¼	3875·9960	220·6974
13/16	3718·9948	216·1813	5/16	3882·8969	220·8937
⅞	3725·7535	216·3777	⅜	3889·8039	221·0901
15/16	3732·5184	216·5748	7/16	3896·7211	221·2864
			½	3903·6343	221·4828
69 in.	3739·2894	216·7704	9/16	3910·5588	221·6791
1/16	3745·8166	216·9667	⅝	3917·4893	221·8755
⅛	3752·8496	217·1631	11/16	3924·4260	222·0718
3/16	3759·6382	217·3594	¾	3931·3687	222·2682
¼	3766·4327	217·5558	13/16	3938·3177	222·4645
5/16	3773·2355	217·7521	⅞	3945·2728	222·6609
⅜	3780·0443	217·9458	15/16	3952·2341	222·8572
7/16	3786·8628	218·1448			
½	3793·6783	218·3412	71 in.	3959·2014	223·0536
9/16	3800·5191	218·5375	1/16	3966·1749	223·2499
⅝	3807·3369	218·7339	⅛	3973·1545	223·4463
11/16	3814·2781	218·9302	3/16	3980·1393	223·6426
¾	3821·0200	219·1266	¼	3987·1301	223·8390
13/16	3827·8708	219·3299	5/16	3994·1292	224·0353
⅞	3834·7277	219·5193	⅜	4001·1344	224·2317
5/16	3841·5908	219·7156	7/16	4008·1447	224·4380

DIAMETERS, AREAS, AND CIRCUMFERENCES.

Dia.	Area.	Circum.	Dia.	Area.	Circum.
1/2	4015·1611	224·6244	73 in.	4185·3966	229·3368
9/16	4022·1837	224·8207	1/16	4192·5665	229·5331
5/8	4029·2124	225·0171	1/8	4199·7424	229·7295
11/16	4036·2473	225·2134	3/16	4206·9230	229·9258
3/4	4043·2882	225·4098	1/4	4214·1107	230·1222
13/16	4050·3354	225·6061	5/16	4221·3027	230·3185
7/8	4057·3886	225·8025	3/8	4228·5077	230·5149
15/16	4064·4481	225·9988	7/16	4235·7109	230·7112
			1/2	4242·9271	230·9076
72 in.	4071·5136	226·1952	9/16	4250·1461	231·1039
1/16	4078·5853	226·3915	5/8	4257·3711	231·3003
1/8	4085·6631	226·5879	11/16	4264·6023	231·4966
3/16	4092·7460	226·7842	3/4	4271·8396	231·6930
1/4	4099·8350	226·9806	13/16	4279·0831	231·8893
5/16	4106·9323	227·1769	7/8	4286·3327	232·0857
3/8	4114·0356	227·3733	15/16	4293·5885	232·2820
7/16	4121·1442	227·5696			
1/2	4128·2587	227·7660	74 in.	4300·8504	232·4784
9/16	4135·3795	227·9623	1/16	4308·1185	232·6747
5/8	4142·5064	228·1587	1/8	4315·3926	232·8711
11/16	4149·6394	228·3550	3/16	4322·1719	233·0674
3/4	4256·7785	228·5514	1/4	4329·9572	233·2638
13/16	4163·9239	228·7477	5/16	4337·2508	233·4601
7/8	4171·0753	228·9441	3/8	4344·5505	233·6565
15/16	4178·2329	229·1401	7/16	4351·8551	233·8528

Diameters, Areas, and Circumferences.

Dia.	Area.	Circum.	Dia.	Area.	Circum.
½	4359·1663	234·0492	76 in.	4536·4704	238·7616
9/16	4366·4835	234·2455	1/16	4543·9333	238·9579
5/8	4373·8067	234·4419	⅛	4551·4023	239·1543
11/16	4381·1361	234·6382	3/16	4558·8794	239·3506
¾	4388·4715	234·8346	¼	4566·3626	339·5470
13/16	4396·3132	235·0309	5/16	4573·8526	239·7433
⅞	4403·1610	235·2273	⅜	4581·3486	239·9397
15/16	4410·5150	235·4236	7/16	4588.8493	240·1360
			½	4596·3571	240·3324
75 in.	4417·8750	235·6200	9/16	4603·8706	240·5287
1/16	4425·2412	235·8163	⅝	4611·3902	240·7251
⅛	4432·6135	236·0127	11/16	4618·9159	240·9214
3/16	4439·9910	236·2090	¾	4626·4477	241·1178
¼	4447·3745	236·4054	13/16	4633·9858	241·3141
5/16	4454·7663	236·6017	⅞	4641·5299	241·5105
⅜	4462·1642	336·7981	15/16	4649·0802	241·7068
7/16	4469·5672	236·9944			
½	4476·9763	237·1908	77 in.	4656·6366	241·9032
9/16	4484·3916	237·3871	1/16	4664·1992	242·0995
⅝	4491·8130	237·5835	⅛	4671·7678	242·2959
11/16	4499·2406	237·7798	3/16	4679·3416	242·4922
¾	4506·6742	237·9762	¼	4586·9215	242·6886
13/16	4514·1141	238·1725	5/16	4694·5097	242·8849
⅞	4521·5600	238·3689	⅜	4702·1039	243·0813
15/16	4528·9622	238·5652	7/16	4709·7033	243·2776

Diameters, Areas, and Circumferences.

Dia.	Area.	Circum.	Dia.	Area.	Circum.
1/2	4717·3087	243·4740	79 in.	4901·6814	248·1864
9/16	4724·9204	243·6703	1/16	4909·4403	248·3827
5/8	4732·5381	243·8667	1/8	4917·2053	248·5791
11/16	4740·1620	244·0630	3/16	4924·9755	248·7754
3/4	4747·7920	244·2594	1/4	4932·7517	248·9718
13/16	4755·8782	244·4557	5/16	4940·5362	249·1681
7/8	4763·0705	244·6521	3/8	4948·3268	249·3645
15/16	4771·1690	244·8484	7/16	4956·1225	249·5608
			1/2	4963·9243	249·7572
78 in.	4778·3736	245·0448	9/16	4971·7319	249·9535
1/16	4786·0344	245·2411	5/8	4979·5456	250·1499
1/8	4793·7012	245·4375	11/16	4987·3663	250·3462
3/16	4801·3732	245·6338	3/4	4995·1930	250·5426
1/4	4809·0512	245·8302	13/16	5003·0316	250·7389
5/16	4817·1375	246·0265	7/8	5910·8642	250·9353
3/8	4824·4299	246·2229	15/16	5018·7091	251·1316
7/16	4832·1275	246·4192			
1/2	4839·8311	246·6156	80 in.	5026·5600	251·3280
9/16	4847·5409	246·8119	1/16	5034·4171	251·5243
5/8	4855·2568	247·0083	1/8	5042·2803	251·7207
11/16	4862·9789	247·2046	3/16	5050·1486	251·9170
3/4	4870·7071	247·4010	1/4	5058·0230	252·1134
13/16	4878·4415	247·5973	5/16	5065·9027	252·3097
7/8	4886·1820	247·7937	3/8	5073·7944	252·5061
15/16	4893·9287	247·9900	7/16	5081·6883	252·7024

Diameters, Areas, and Circumferences.

Dia.	Area.	Circum.	Dia.	Area.	Circum.
½	5089·5883	252·8988	82 in.	5281·0296	257·6112
9/16	5097·4941	253·0951	1/16	5289·0781	257·8075
⅝	5105·4060	253·2915	⅛	5297·1426	258·0039
11/16	5113·8248	253·4878	3/16	5305·2073	258·2002
¾	5121·2497	253·6842	¼	5313·2780	258·3966
13/16	5129·1855	253·8805	5/16	5321·3570	258·5929
⅞	5137·1173	254·0769	⅜	5329·4421	258·7893
15/16	5145·0603	254·2732	7/16	5337·5324	258·9856
			½	5345·6287	259·1820
81 in.	5153·0094	254·4696	9/16	5353·7809	259·3783
1/16	5160·9647	254·6659	⅝	5361·8391	259·5747
⅛	5168·9260	254·8623	11/16	5369·9543	259·7710
3/16	5176·8925	255·0586	¾	5378·0755	259·9674
¼	5184·8651	255·2550	13/16	5386·2026	260·1637
5/16	5192·8460	255·4513	⅞	5394·3358	260·3601
⅜	5200·8329	255·6477	15/16	5402·4552	260·5564
7/16	5208·8250	255·8440			
½	5216·8231	256·0404	83 in.	5410·6206	260·7528
9/16	5224·8271	256·2367	1/16	5418·7722	260·9491
⅝	5232·8371	256·4331	⅛	5426·9299	261·1455
11/16	5240·8568	256·6294	3/16	5435·0928	261·3418
¾	5248·8772	256·8258	¼	5443·2617	261·5382
13/16	5256·9061	257·0221	5/16	5451·4389	261·7345
⅞	5264·9411	257·2104	⅜	5459·6222	261·9309
15/16	5272·9828	257·4148	7/16	5467·8106	262·1272

Diameters, Areas, and Circumferences.

Dia.	Area.	Circum.	Dia.	Area.	Circum.
½	5476·0051	262·3236	85 in.	5674·5150	267·0360
9/16	5484·2054	262·5199	1/16	5682·8630	267·2323
⅝	5492·4118	262·7163	⅛	5691·2170	267·4287
11/16	5500·6252	262·9126	3/16	5699·5762	267·6250
¾	5508·8446	263·1090	¼	5707·9415	267·8214
13/16	5517·0699	263·3053	5/16	5716·3151	268·0177
⅞	5525·3012	263·5017	⅜	5724·6947	268·2141
15/16	5533·5388	263·6980	7/16	5733·0795	268·4104
			½	5741·4703	268·6068
84 in.	5541·7824	263·8944	9/16	5749·8670	268·8031
1/16	5550·0322	264·0907	⅝	5758·2697	268·9997
⅛	5558·2881	264·2871	11/16	5766·6794	269·1958
3/16	5566·5491	264·4834	¾	5775·0952	269·3922
¼	5574·8162	264·6798	13/16	5783·5168	269·5885
5/16	5583·0916	264·8761	⅞	5791·9445	269·7849
⅜	5591·3730	265·0725	15/16	5800·3784	269·9812
7/16	5599·6596	265·2688			
½	5607·9523	265·4652	86 in.	5808·8184	270·1776
9/16	5616·2508	265·6615	1/16	5817·2651	270·3739
⅝	5624·5554	265·8579	⅛	5825·7168	270·5703
11/16	5632·8662	266·0542	3/16	5834·1742	270·7666
¾	5641·1845	266·2506	¼	5842·6376	270·9630
13/16	5649·5071	266·4469	5/16	5851·1093	271·1593
⅞	5657·8357	266·6433	⅜	5859·5871	271·3557
15/16	5666·1723	266·8396	7/16	5868·0701	271·5520

Diameters, Areas, and Circumferences.

Dia.	Area.	Circum.	Dia.	Area.	Circum.
1/2	5876·5591	271·7484	88 in.	6082·1376	276·4608
9/16	5885·0540	271·9447	1/16	6090·7801	276·6671
5/8	5893·5549	272·1411	1/8	6099·4287	276·8535
11/16	5902·0620	272·3374	3/16	6108·0824	277·0498
3/4	5910·5767	272·5338	1/4	6116·7422	277·2462
13/16	5919·0965	272·7301	5/16	6125·4103	277·4425
7/8	5927·6224	272·9265	3/8	6134·0844	277·6389
15/16	5936·1545	273·1228	7/16	6144·2637	277·8352
			1/2	6151·4491	278·0316
87 in.	5944·6926	273·3192	9/16	6160·1403	278·2279
1/16	5953·2369	273·5155	5/8	6169·8376	278·4243
1/8	5961·7873	273·7119	11/16	6177·5418	278·6206
3/16	5970·3429	273·9082	3/4	6186·2521	278·8170
1/4	5978·9045	274·1046	13/16	6194·9683	279·0133
5/16	5987·4749	274·3009	7/8	6203·6905	279·2097
3/8	5996·0504	274·4973	15/16	6212·4189	279·4060
7/16	6004·6315	274·6936			
1/2	6013·2187	274·8900	89 in.	6221·1534	279·6024
9/16	6021·8117	275·0863	1/16	6229·8941	279·7987
5/8	6030·4108	275·2827	1/8	6238·6408	279·9951
11/16	6039·0169	275·4790	3/16	6247·3927	280·1914
3/4	6047·6290	275·6754	1/4	6256·1507	280·3878
13/16	6056·2470	275·8717	5/16	6264·9170	280·5841
7/8	6064·8710	276·0681	3/8	6273·6893	280·7805
15/16	6073·5013	276·2644	7/16	6282·4668	280·9768

Diameters, Areas, and Circumferences.

Dia.	Area.	Circum.	Dia.	Area.	Circum.
$\frac{1}{2}$	6291·2503	281·1732	91 in.	6503·8974	285·8856
$\frac{9}{16}$	6300·0397	281·3695	$\frac{1}{16}$	6512·8344	286·0819
$\frac{5}{8}$	6308·8351	281·5659	$\frac{1}{8}$	6521·7775	286·2783
$\frac{11}{16}$	6317·6375	281·7622	$\frac{3}{16}$	6530·7258	286·4746
$\frac{3}{4}$	6326·4460	281·9586	$\frac{1}{4}$	6539·6801	286·6710
$\frac{13}{16}$	6335·2603	282·1549	$\frac{5}{16}$	6548·6427	286·8673
$\frac{7}{8}$	6344·0807	282·3513	$\frac{3}{8}$	6557·6114	287·0637
$\frac{15}{16}$	6352·9073	282·5476	$\frac{7}{16}$	6566·5857	287·2600
			$\frac{1}{2}$	6573·5651	287·4564
90 in.	6361·7400	282·7440	$\frac{9}{16}$	6584·5511	287·6527
$\frac{1}{16}$	6370·5789	282·9403	$\frac{5}{8}$	6593·5431	287·8491
$\frac{1}{8}$	6379·4238	283·1367	$\frac{11}{16}$	6602·5443	288·0454
$\frac{3}{16}$	6388·7739	283·3330	$\frac{3}{4}$	6611·5462	288·2418
$\frac{1}{4}$	6397·1300	283·5294	$\frac{13}{16}$	6620·5569	288·4381
$\frac{5}{16}$	6405·9944	283·7257	$\frac{7}{8}$	6629·5736	288·6345
$\frac{3}{8}$	6414·8649	283·9221	$\frac{15}{16}$	6638·5967	288·8388
$\frac{7}{16}$	6423·7906	284·1184			
$\frac{1}{2}$	6432·6223	284·3148	92 in.	6647·6258	289·0272
$\frac{9}{16}$	6441·5101	284·5111	$\frac{1}{16}$	6656·6609	289·2235
$\frac{5}{8}$	6450·4039	284·7075	$\frac{1}{8}$	6665·7021	289·4199
$\frac{11}{16}$	6459·3043	284·9038	$\frac{3}{16}$	6674·7485	289·6162
$\frac{3}{4}$	6468·2107	285·1002	$\frac{1}{4}$	6683·8010	289·8125
$\frac{13}{16}$	6477·1232	285·2965	$\frac{5}{16}$	6692·8618	290 0089
$\frac{7}{8}$	6486·0418	285·4929	$\frac{3}{8}$	6701·9286	290·2053
$\frac{15}{16}$	6494·9566	285·6892	$\frac{7}{16}$	6711·5001	290·4016

Diameters, Areas, and Circumferences.

Dia.	Area.	Circum.	Dia.	Area.	Circum.
$\frac{1}{2}$	6720·0787	290·5980	94 in.	6939·7946	295·3194
$\frac{9}{16}$	6729·6628	290·7943	$\frac{1}{16}$	6949·5261	295·5067
$\frac{5}{8}$	6738·2530	290·9907	$\frac{1}{8}$	6958·2636	295·7031
$\frac{11}{16}$	6747·3497	291·1870	$\frac{3}{16}$	6968·0064	295·8994
$\frac{3}{4}$	6756·4525	291·3834	$\frac{1}{4}$	6976·7552	296·0958
$\frac{13}{16}$	6765·5614	291·5797	$\frac{5}{16}$	6986·0123	296·2921
$\frac{7}{8}$	6774·6763	291·7761	$\frac{3}{8}$	6995·2755	296·4885
$\frac{15}{16}$	6783·7975	291·9724	$\frac{7}{16}$	7004·5439	296·6848
			$\frac{1}{2}$	7013·8183	296·8812
93 in.	6792·9248	292·1688	$\frac{9}{16}$	7023·0988	297·0775
$\frac{1}{16}$	6802·0581	292·3651	$\frac{5}{8}$	7032·3853	297·2739
$\frac{1}{8}$	6811·1974	292·5615	$\frac{11}{16}$	7041·6784	297·4702
$\frac{3}{16}$	6820·3420	292·7578	$\frac{3}{4}$	7050·9775	297·6666
$\frac{1}{4}$	6829·4927	292·9542	$\frac{13}{16}$	7060·2827	297·8629
$\frac{5}{16}$	6838·6517	293·1505	$\frac{7}{8}$	7069·5940	298·0593
$\frac{3}{8}$	6847·8167	293·3469	$\frac{15}{16}$	7075·9116	298·2556
$\frac{7}{16}$	6856·9869	293·5432			
$\frac{1}{2}$	6866·1631	293·7396	95 in.	7088·2352	298·4520
$\frac{9}{16}$	6875·3454	293·9359	$\frac{1}{16}$	7097·5738	268·6483
$\frac{5}{8}$	6884·5338	294·1323	$\frac{1}{8}$	7106·9005	298·8447
$\frac{11}{16}$	6893·7337	294·3286	$\frac{3}{16}$	7116·7415	299·0400
$\frac{3}{4}$	6902·9296	294·5350	$\frac{1}{4}$	7125·5885	299·2374
$\frac{13}{16}$	6912·1366	294·7213	$\frac{5}{16}$	7134·9443	299·4337
$\frac{7}{8}$	6921·3497	294·9177	$\frac{3}{8}$	7144·3052	299·6301
$\frac{15}{16}$	6930·5691	295·1140	$\frac{7}{16}$	7153·6717	299·8264

Diameters, Areas, and Circumferences.

Dia.	Area.	Circum.	Dia.	Area.	Circum.
½	7163·0443	300·0228	97 in.	7389·8288	304·7352
9/16	7172·4230	300·2191	1/16	7399·3548	304·9315
⅝	7181·8077	300·4155	⅛	7408·8868	305·1279
11/16	7191·1989	300·6118	3/16	7418·6241	305·3242
¾	7200·5962	300·8082	¼	7427·9675	305·5206
13/16	7209·9096	301·0045	5/16	7437·5192	305·7169
⅞	7219·4090	301·2009	⅜	7447·0769	305·9133
15/16	7228·8248	301·3972	7/16	7456·6398	306·1096
			½	7466·2087	306·3060
96 in.	7238·2466	301·5936	9/16	7475·7837	306·5023
1/16	7247·6741	301·7899	⅝	7485·3648	306·6987
⅛	7257·1083	301·9863	11/16	7494·9524	306·8950
3/16	7266·5474	302·1826	¾	7504·5460	307·0914
¼	7275·9926	302·3790	13/16	7514·1457	307·2877
5/16	7285·4461	302·5753	⅞	7523·7515	307·4841
⅜	7294·9056	302·7717	15/16	7533·3686	307·6804
7/16	7304·3703	302·9680			
½	7313·8411	303·1644	98 in.	7542·9818	307·8768
9/16	7323·3179	303·3607	1/16	7552·6060	308·0731
⅝	7332·8008	303·5571	⅛	7562·2362	308·2695
11/16	7342·2902	303·7534	3/16	7575·8717	308·4658
¾	7351·7857	303·9498	¼	7581·5132	308·6622
13/16	7361·2873	304·1461	5/16	7591·1630	308·8585
⅞	7370·7949	304·3425	⅜	7600·8189	309·0549
15/16	7380·3088	304·5388	7/16	7610·4800	309·2512

Diameters, Areas, and Circumferences

Dia.	Area.	Circum.
½	7620·1471	309·4476
9/16	7629·8203	309·6439
5/8	7639·4995	309·8403
11/16	7649·1853	310·0366
¾	7658·8771	310·2330
13/16	7668·5750	310·4293
7/8	7678·2790	310·6257
15/16	7687·9893	310·8220
99 in.	7697·7056	311·0184
1/16	7707·4279	311·2147
1/8	7717·1563	311·4111
3/16	7726·8900	311·6074
¼	7736·6297	311·8038
5/16	7746·3777	312·0001
3/8	7756·1318	312·1965
7/16	7765·8910	312·3928
½	7775·6563	312·5892
9/16	7785·4277	312·7855
5/8	7795·2051	312·9819
11/16	7804·9890	313·0782
¾	7814·7790	313·3746
13/16	7824·5751	313·5709
7/8	7834·3772	313·7673
15/16	7844·1856	313·9636

Dia.	Area.	Circum.
100 in	7854·0000	314·1600
¼	7893·3190	314·9454
½	7932·7360	315·7308
¾	7972·2120	316·5162
101 in	8011·8652	317·3016
¼	8051·5772	318·0870
½	8091·3870	318·8724
¾	8131·2953	319·6578
102 in	8171·3016	320·4432
¼	8211·4060	321·2286
½	8251·6084	322·0140
¾	8291·8696	322·7994
103 in	8332·3085	323·5848
¼	8372·8056	324·3702
½	8413·4008	325·1556
¾	8454·0944	325·9410
104 in	8494·8864	326·7264
¼	8535·7760	327·5118
½	8576·7640	328·2972
¾	8617·8504	329·0826

DIAMETERS, AREAS, AND CIRCUMFERENCES.

Dia.	Area.	Circum.	Dia.	Area.	Circum.
105 in	8659·0348	329·8680	108 in	9160·9056	339·2928
¼	8700·3176	330·6534	½	9245·9248	340·8636
½	8741·6980	331·4388			
¾	8783·1772	332·2242			
			109 in	9331·3372	342·4344
106 in	8824·7544	333·0096	½	9417·1420	344·0052
½	8908·2028	334·5804			
107 in	8992·0444	336·1512	110 in	9503·3400	345·5760
½	9076·2784	337·7220			

SUBJECT INDEX.

GENERAL INDEX.

Alphabetical Lists of the prepaid Subscribers to the second Edition of this Work.

Civil and Consulting Engineers.

	No.
Adams	1
Allan	1
Baker	1
Bradford	1
Bamber	1
Balfour Tyson & Co.	3
Bateman Latrobe	1
Barry, Jones, & Co. (Ventilating Engineers)	1
Beauchamp Tower	1
Bennett	1
Brereton	1
Beldam	3
Bewick	1
Bessemer, Sir H.	2
Blakesley	1
Boyd	1
Brunton	1
Browne	1
Brunlees	1
Bridges	1
Bruce	1
Brougham	1
Brown, Oswald	1
Browne	1
Bryan	3
Bromfield	1
Birch	1
Bush	1
Campbell Evans	1
Crampton	1
Craven	1
Claudet	1
Coxon	1
Clinkskill	2
Cleminson	1
Currey	1
Corry	1
Coode, Sir John	1
Church, Jabez	1
Daniel	1
Darlington	6
Davis	1
Douglas, Sir James	1
Dudgeon	1
Drower	1
Dick (Delta Metal)	1
Drake	1
Donaldson	1
Eckersley	1
Edinburgh, Duke of	1
Ellison	1
Eggleton	1
Edwards	1

	No.
Evans Campbell ..	1
Fairlie	1
Flower	2
Field	1
Feld	1
Finch	1
(Sanitary Engineer)	
Fox	1
Fox Mackinson ..	1
Fyson	1
Firby	1
Fforde	1
Furness	1
Forbes	1
Gray	1
Godfrey	1
Greathead	1
Greig	1
Gould	1
Gulland	1
Gorham	1
Hawkshaw, Sir John	1
Hawksley	1
Hastins	1
Hawley	1
Hamand	1
Hartley, Sir Charles ..	1
Harrison	1
(Consulting Fire Engineer.)	
Holtham	1
Hemel	1
Hopkins	1
Hopkinson, Dr., F.R.S.	1
Homersham	2

	No.
Hughes	1
Harfield	1
Hald	4
Haughton	1
Horn	1
Hutton Vignoles ..	1
Homan Rogers ..	1
Indian E. R... ..	1
Jacob	1
Jacohs	1
Jackson	2
Jenkins	1
Kinniple & Morris	1
Kingsbury	1
Lennox	1
Lockhart	1
Livesey	1
Law	1
Lucas Brothers ..	1
Lemon	1
Lineff & Jones .	1
Lewis	1
Lowtham	1
Marley, Pinchin and Marley	1
MacNeill, Hotchkiss, and Co.	1
MacIntyre	1
Mackimmion .	1
Mathewson	1
Miken	1
Mellis	1
Mansergh	1
McLaren	1

	No.
More	1
Macintosh	1
Moffatt	1
Maynard	1
Maudesly	1
Myall	1
Menzies & Blagburn	1
Mathews	1
Morant	1
Norbery .. .	1
Neate	1
Ogilvie	1
Ormsby	1
Phipson Wilson ..	1
Prim	1
Price Williams ..	1
Punchard	1
Provis, Wilson ..	1
Park	1
Perry	1
Quick & Sons ..	1
Ridley Noel	1
Redfern	1
Robinson	1
Roberts	1
Rofe	1
Reilly	1
Reid	1
Rickard	1
Rickards	1
Robertson	1
Serjeant	50
Somerville	1
Smith	2
Sanbergue de .	1
Sage	1
Sadler	1
Slade	1
Stockman	1
Stanger	1
Simpson Telford ..	1
Spiers	1
Stuart	1
Smith	1
Stephenson Gurdon ..	1
Stewart	1
Steel, Young & Co. ..	1
Sturgeon	1
Sinclair	1
Stoney	1
Surson	1
Thomas	1
Tweddell	1
Tindal Atkinson ..	1
Thornton	1
Thuey	1
Tahourdin	1
Temple	1
Urquhart	1
Vernon Harcourt ..	1
Viger	1
Valentine	1
Winter	1
Walton	1
Wakefield	1
Walker	2

	No.		No.
Warburton	1	Willie	1
Wedekind Herman ..	1	Walmisley	1
Wilson	1	Walton	1
Williamson Jarvis ..	1	Woods	1
Wilberg	1	Wigner & Harland ..	1
Walker	1	Walker	1
Wylie & Fulton ..	1	Yerburgh	1

Electrical Engineers.

	No.		No.
Albright	1	Reskenzaun	1
Capito	1	Phillips & Harrison ..	1
Blackburn	1	Goodwin & How ..	1
Crompton	2	Swete & Main ..	3
Francis	1	Wilger	1
Raworth	1		

Gas Engineers.

	No.		No.
Braidwood	1	Packham	1
Birkett	1	Lacey	1
Beale	1	M'Minn	1
Carpenter	1	M'Minn	1
Delatouche	1	Morris	1
Jago	1	Somerville	1

Locomotive Engineers.

	No.		No.
Adams	1	Levett	1
Ellis	1	Matthews	1
Crampton	1	Otway	1
Fairlie	1	Sant	1
Gould	1	Spence	1
Holden	2	Satchell	1
Jacob	1	Tomlinson	1
Kennedy	1	Trevithick	1
Kirtley	1	Worsdell	1

Marine and Mechanical Engineers.

	No.
Adamson	6
Adams & Co.	1
Anderson	1
Anderson & Galleway	1
Appleby	1
Alliman	1
Aland	1
Applegarth (Diving Apparatus)	1
Anderson	1
Bastin & Lanson	1
Barnet	1
Batting	1
Baynes	1
Burke	1
Bellatt	1
Bramham	1
Balfour, J.	1
Blundell	1
Brophy	1
Boaz	1
Ball	1
Baillie	1
Booth	1
Botten	1
Butterfield & Co.	1
Campbell (Supert. B. I. S. N. Co.)	1
Chapman	2
Cornes	1
Cornish (Mining Engineer)	1
Course	1
Craig	1
Churchill	1
Chaplin	1
Coley	1
Cousins	1
Corney	1
Chater	1
Cayzer	1
Cloak	1
Chaden	1
Christianson	1
Donkin	1
Darke	1
Death	2
Davison	1
Davey	1
Davisons	1
Dawson	1
Davis	1
Davis	1
Dewrance	2
Dean	1
Donkin	1
Dixie	1
Davis	1
Davy, Paxman & Co.	1
Duncan	1
Edwards	1
Elliott	1

	No.
Esson	1
Edmonds	1
Everitt	1
Frielday	1
Farman	1
Fraser..	1
Fardon	1
Fletcher	1
Fuller..	1
Fricker	1
Goodfellow	1
Gardner	1
George .. .. (Engineering Draughtsman)	1
Gillispie	1
Gray	1
Gordon .. .	1
Green	1
Glover	1
Grover	1
Gwynne	1
Gillman & Spencer ..	1
Gorman	1
Harris	1
Harris..	1
Harris..	1
Harris..	2
Harrison	1
Haskins	2
Harvey	1
Haskins	1
Hayes	1
Heslop	1
Humble	1

	No.
Hutchinson	1
Hindley	1
Hind	1
Holman	1
Hopkins	1
Hopkins	1
Hanly	1
Howard	1
Harman	1
Hale, Ayle & Hale ..	1
Jenner	1
Jeakes..	2
Jones	1
Jones	1
Kirkaldy & Son ..	1
Kuhleman	1
Knowles	1
Ladd	1
Leber	1
Lewis	1
Lloyd's Register Office	1
Le June	1
Love	1
London and Colonial Engineering Co. ..	1
Maudsley, Sons & Field	
Mc'Intyre	6
Martin	1
Martin	1
Maclesend	1
Manuel .. .. (Supert. P. & O. Co.)	1
Macdonald	1

	No.
Masters .. .. (Supert. G. S. N. Co.)	1
Moreland	3
Marshall	1
More	1
Mindsore	2
Mirrlees	1
Manlove & Alliott ..	1
Mackay	1
Moss	1
Masson, Scott & Bertram	1
Mountain	1
Mowbray	1
Newan	1
Nilehay	1
Owen	1
Parker	1
Parkes	1
Pearce	4
Penn, J. & Son's ..	6
Potter	1
Proctor .. .. (Agricultural Digger.)	1
Pepper Mill Company	1
Pontifex & Wood ..	1
Ram	1
Reid	1
Ransome, Joselyn & Co.	1
Rennie	2
Ronald	1
Robinson	1
Robinson	1
Selwyn, Admiral ..	1
Sharer . .. (Engineer's Store Dealer)	
Smith & Co.	1
Stevenson & Davis .. (Patent "Cestus" Vertical Boiler)	1
Simpson	2
Spurr	1
Scot-Russell	1
Southgate Engineering Co.	1
Thompson	1
Tilley & Sons .. (Well Engineers)	
Voss	1
Wright	2
Westwood & Bailie ..	4
West	2
Wedeking	1
Whimshurst, Hollick & Co.	1
Walker	1
White	1
Whealey	1
Wheatley, Kirk & Co.	1
Whitman	1
Wilson	1
Wingate	1
Young	1

MARINE ENGINEERS (Afloat) in the Main "Lines." Port of London.

One Copy Each.

Adamson, E. P.
Allison, A.
Adam, R.
Ashmore, W.
Auld, J.
Anderson, J. M.
Allen, F.
Aldridge, T.
Ameida, F. de
Beach, T. W.
Beldam, J.
Banks, J. J,
Brown, R. P.
Blane, A.
Braidwood, J.
Bibby, R.
Bovey, J. & R.
Bullock, R.
Brownfield, J.
Balfour, R.
Burrows, J.
Brown, W.
Bambright, W.
Baxter, P.
Buddy, J.
Boyd, P.
Batchelor, J. G.
Bosestow, G.
Buckham. J.
Borland, A.
Brock, A.
Bamringer, H.
Barclay, R. M.
Bethel, E.
Boyle, R.
Bullock A.
Brown, D.
Baker, W. H.
Casse, J.
Campbell, J.
Croft, J.
Cross, R. J.
Carmichael, D.
Campbell, D.
Cooper, W.
Cornell, E.
Collier, D.
Coults, J.
Croal, G.
Cowper, D. L.
Cleghorn, C.
Chuck, A.

Campbell, C.
Carlton, T.
Cameron, D.
Cranck, J.
Darley, J.
Duncan, D.
Dawes, W. H.
Davidson, W.
Donald, P.
Duguid, J.
Dunn, D.
Dibb, J.
Ditchburn, J.
Davis, S.
Dawson, S.
Darry, W. H.
Edwards, A.
Elder, A.
Eyre, H.
Ellans, R.
Essens, L. Van.
Ewins, H. W.
Fasse, J.
Ferris, P.
Findlay, M.
Fairburn, J.
Fyfe, R.
Fraser, G.
Ferguson, U.
Ferguson, A. S.
Fairweather, F.
Fenton, U.
Fletcher, A.
Gillispie, D.
Girvin, J.
Gondie, H.
Gumbell, F.
Grigg, D.
Gauldie, R. L.
Grayston, W.
Graham, J. G.
Grigg, J.
Gardiner, W.
Grant, J.
Gilmour, J.
Granger, G.
Gattrell, M.
Gray, J.
Geary, J.
George, R.
Gifford, W. H.
Gondon, W.
Gilchrist, P. B.
Gunson, W. B.
Highet, H.
Hogert, R.
Hosking, F. A.
Henderson, D.
Hyde, R. E.
Highet, J.
Hall, S. E.
Hill, W.
Higginson, R.
Hiron, J.
Halfyard, P.
Harper, G.
Harper, J. H.
Hunter, J.

Hayes, S. A.
Hobson, A. W.
Hutton, D. B.
Hague, J.
Harding, J. E.
Howe, G.
James, W. F.
Johnson, J.
Jones, G.
Jago, W. R.
Kidd, A.
Kaye, A.
Kelso, J.
Kinley, W.
Kerr, J.
Keladeta, N.
Kingsworth, G. J.
Kay, J. R.
Lawrie, T
Lee, J. J.
Leburn, W.
Lamont, T. W.
Lamb, W.
Lome, H.
Linklater, H.
Latchford, J.
Love, W.
Leabrook, H. T.
Mackie, A.
Marten, W.
Martyn, W. C.
Maher, M.
Macdonald, C. A.
Milne, A.
McLiam, J.
McKinnon, D.
Meade, P.
MacKongee, L.
McAllister, W.
McAllan, A.
McMurhciey, J.
McColl, J.
McKinnon, D.
McEwan, H. D.
Murray, W. G. D.
Morrison, A.
McMask, P. C.
Muirhead, J. S.
Montgomery, W. A.
Macker, A.
Malcolm, J.
Mackay, A. B.
Morgan, G. D.
MacLaughlin, J.
Millar, G.
McIntyre, J.
McKinnon, J.
Morgan, J. T.
Murray, J. T.
Moss, T.
McMurchiey.
McIndiarn.
Noble, C.
Neill, J.
Oswald, C.
Ould, J. G.
Organ, W. J.
Paton, W.

Prahm, A. C. C.
Purvis, W. H.
Power, C. S.
Philpott, E.
Paterson, G.
Phillips, J.
Pattison, J.
Phillips, E. C.
Peacock, J. E.
Pearce, R.
Puem, J. J.
Pulton, J.
Quesne, D. Le
Randall, J. W.
Robertson, W.
Riddell, R.
Rogers, W. H.
Ross, E. W.
Ross, W. M.
Rennie, D. C.
Robertson, E.
Schurr, A. E.
Rippard, S.
Reid, J.
Reia, D.
Randall, M.
Rendle, A. F.
Russell, R.
Reid, R.
Robertson, A.
Smith, A.
Stevenson, J. B.
Stephen, J. W.
Smith, J. F.
Stoddart, J.
Smith, S.
Stephens, G.
Syme, S. A.
Scott, R.
Spence, T. B.
Sara, E.
Smith, D.
Sim, D. A.
Stewart, J.
Slater, C.
Smart, O. G.
Struthers, A.
Stocks, J. K.
Swinton, A. C.
Simpson, J. M.
Stephens, J.
Smith, W. P.
Smith, T. S.
Scoukar, R.
Shearer, G.
Sellex, J. W.
Small, H. C.
Sinclair, J.
Sturrock, J.
Scott, J.
Stevenson, W.
Stephens, J.
Turnbull, R.
Thaw, W.
Taylor, A.
Thompson, J. H.
Thomson, T.
Thomas, T. C.

Truscott, E.
Tricker, C. H.
Thomas, W.
Turnbull, M.
Thompson, J. H.
Thomson, D.
Thomson, B.
Thomson, W.
Trotman, T. H.
Todd, C.
Trevelyan, C. W.
Taylor, C.
Urquhart, H.
Wilson, W. W.
Willis, F. R. T.
Warner, J.
Watters, H.
Watt, H.
Wordall, R.
Wilkes,
Williamson, J.
Watt, W.
Walker, M.
Walker, W. C.
Williams, W. J.
Williamson, D.
White, A.
White, W.
Wilkinson, J.
Woods, J.
Walker, M.
Young, J.

Patent Agents.

One Copy Each.

Andrew and Co.
Allison Brothers
Browne and Co.
Browne
Coxhead
Cumberpatch
Downing
Edwards and Co.
Fell and Wilding
Gardner
Gardner Cotton
Gedge
Hodges and Russ
Johnson
Jensen
Messer and Thorpe
Mewburn
Rogers
Spence and Son
Thompson and Boult
Whiteman
Wilson

WATERWORKS ENGINEERS.

Beeson	1	Hoskins	1
Carruthers	1	Lynsea	1
Daires	1	Loam	1
Frazer	1	Loam	2
Francis	1	Morris	1
Goochman	1	Robinson	1
George	1	Restler	1
Hack	1	Taylor and Sons	1
Hack	1	Trott	1

YACHT CONSULTING ENGINEERS.

One Copy Each.

Barnaby
Brown
Christie
Douglas
Ensom
Hewitt
Holdsworth
Holdsworth.
Miller and Tupp
Storey
Thomson
Thornycroft
Westwood
Wingfield

Lightning Source UK Ltd.
Milton Keynes UK
UKHW021437191118
332600UK00012B/1205/P